Touchless Fingerprint Biometrics

Touchless Fingerprint Biometrics

Ruggero Donida Labati
Vincenzo Piuri
Fabio Scotti

CRC Press
Taylor & Francis Group
Boca Raton London New York

CRC Press is an imprint of the
Taylor & Francis Group, an **informa** business

CRC Press
Taylor & Francis Group
6000 Broken Sound Parkway NW, Suite 300
Boca Raton, FL 33487-2742

First issued in hardback 2020

© 2016 by Taylor & Francis Group, LLC
CRC Press is an imprint of Taylor & Francis Group, an Informa business

No claim to original U.S. Government works

ISBN-13: 978-1-4987-0761-9 (hbk)

Library of Congress Cataloging-in-Publication Data

Labati, Ruggero Donida.
  Touchless fingerprint biometrics / Ruggero Donida Labati, Vincenzo Piuri, and Fabio Scotti.
    pages cm. -- (Series in security, privacy, and trust)
  Includes bibliographical references and index.
  ISBN 978-1-4987-0761-9 (alk. paper)
  1. Biometric identification. 2. Fingerprints. I. Piuri, Vincenzo. II. Scotti, Fabio. III. Title.

TK7882.B56L33 2016
006.4--dc23   2015017253

Visit the Taylor & Francis Web site at
http://www.taylorandfrancis.com

and the CRC Press Web site at
http://www.crcpress.com

Contents

Preface

Biometric systems recognize individuals by analyzing biological or behavioral characteristics, called biometric traits. These systems are widely used in security scenarios, and their diffusion is constantly increasing in governmental, industrial, and consumer applications. The most accurate biometric technologies (e.g., touch-based fingerprint recognition systems and iris-based systems) require a high level of user cooperation to acquire samples of the considered biometric trait.

Touchless fingerprint recognition is a recent set of techniques designed to reduce the acquisition constraints of those biometric technologies, while featuring high recognition accuracy. Touchless approaches allow for increasing the usability and reduce the hardware costs with respect to traditional fingerprint recognition systems. Typical applications can be in access control to high security areas and systems, border control, and consumer systems based on cameras integrated in smartphones or tablets.

This book aims to offer the first comprehensive analysis of touchless fingerprint recognition technologies: The state of the art is reviewed, industrial applications are described, and original technologies are presented in this rapidly moving field. Traditional methods and original research studies on two-dimensional and three-dimensional biometric systems are summarized and compared, encompassing all the steps of the biometric recognition process, with specific reference to acquisition, quality assessment, computation of touch-equivalent images, compensation of distortions, matching, and generation of synthetic samples. Hardware and software solutions are analyzed for key problems such as non-idealities of touchless images (e.g., reflections, shadows, and complex background), the presence of distortions due to perspective effects, uncontrolled finger placement, inconstant resolution of the ridge pattern, and reconstruction and processing of three-dimensional models.

Beginners in biometric technologies and their applications can start profitably from the initial introductive chapters, while advanced readers can find specific material in the rest of the book where all modules composing a touchless finger-print system are described. Professionals can find inspiration for using biometric touchless fingerprint recognition in a variety of applications, including innovative solutions especially conceived for non-highly critical applications, for example, on smart phones and tablets and in the Internet-of-Things.

Acknowledgments

This work was partially funded by the following organizations:

- The European Commission under the project "ABC4EU" (contract no. FP7-312797)
- The Italian Ministry of Research under the project "PRIN" "GenData 2020" (contract no. 2010RTFWBH)
- The EU under the 7FP project "PrimeLife" (contract no. 216483)
- The Italian Ministry of Education, Universities, and Research (MIUR) under the project "Priv-Ware" (contract no. 2007JXH7ET)

The cover of this book has been created and drafted by Laura Vanelli Tagliacane.

Authors

Ruggero Donida Labati (PhD in computer science, 2013) is a postdoctoral research assistant in computer science at the Università degli Studi di Milano, Italy. He has been a visiting researcher at Michigan State University, USA.

His main research interests are biometric systems, biometric encryption and privacy-compliant biometric templates, signal and image processing, computational intelligence algorithms, and industrial applications. Original results have been published in over 30 papers in international journals, proceedings of international conferences, and book chapters. He has participated in several national and international research projects involving different aspects of signal and image processing, pattern recognition, and computational intelligence for biometrics and industrial applications.

He is a secretary of the IEEE (Institute of Electrical and Electronics Engineers) Italy Section Computational Intelligence Society Chapter since 2013. He has been program co-chair of the 2015 IEEE International Conference on Computational Intelligence and Virtual Environments for Measurement Systems and Applications and the 2014 IEEE Workshop on Environmental Energy and Structural Monitoring Systems. He is and has been a member of the technical program committee of more than 50 international conferences. He is an IEEE member.

 **Vincenzo Piuri** (PhD in computer engineering, 1989) is full professor in computer engineering at the Università degli Studi di Milano, Italy. He has been associate professor at Politecnico di Milano, Italy, and visiting professor at the University of Texas at Austin and at George Mason University, USA. He has been director of the Department of Information Technology at the Università degli Studi di Milano, Italy.

His main research interests are biometrics, pattern analysis and recognition, signal and image processing, theory and industrial applications of neural networks, machine learning, intelligent measurement systems, industrial applications, fault tolerance, digital processing architectures, embedded systems, and arithmetic architectures. Original results have been published in over 350 papers in international journals, proceedings of international conferences, books, and book chapters. He participated in several national and international research projects involving different aspects of signal and image processing, pattern recognition, and computational intelligence for biometrics and industrial applications.

He is IEEE vice president for Technical Activities (2015) and IEEE director. He has been IEEE delegate for Division X, president of the IEEE Computational Intelligence Society, vice president for publications of the IEEE Instrumentation and Measurement Society and the IEEE Systems Council, vice president for membership of the IEEE Computational Intelligence Society, vice president for education of the IEEE Biometrics Council, chair of the IEEE Biometrics Certification Committee, and vice-chair of the IEEE TAB NTDC Technical Committee on Biometrics.

He has been general chair or program chair in 50+ international conferences and workshops, mainly sponsored by IEEE and ACM (Association for Computing Machinery), in the areas of computational intelligence, measurement systems, and biometrics. He is editor-in-chief of the *IEEE Systems Journal* (2013–2015), and has been associate editor of the *IEEE Transactions on Neural Networks* and the *IEEE Transactions on Instrumentation and Measurement*.

He is Fellow of the IEEE (2001), distinguished scientist of ACM (2008), and senior member of INNS (International Neural Network Society) (2009). He received the IEEE Instrumentation and Measurement Society Technical Award (2002) for the contributions to the advancement of theory and practice of computational intelligence in measurement systems and industrial applications, the IEEE Instrumentation and Measurement Society Distinguished Service Award (2008), and the IEEE Computational Intelligence Society Meritorious Service Award (2009). He is distinguished lecturer for the IEEE Systems Council since 2013 and distinguished lecturer of the IEEE Computational Intelligence Society since 2014. He is honorary professor at Obuda University, Budapest, Hungary since 2014, and guest professor (equivalent to honorary professor) at the Guangdong University of Petrochemical Technology, China since 2014.

Fabio Scotti (PhD in computer engineering, 2003) is associate professor in computer science at the Università degli Studi di Milano, Italy.

His research interests include biometric systems, biometric encryption and privacy-compliant biometric templates, multimodal biometric systems, signal and image processing, computational intelligence algorithms, industrial applications, high-level system design. Original results have been published in over 90 international journals, proceedings of international conferences, and book chapters. He has participated in several national and international research projects involving different aspects of signal and image processing, pattern recognition, and computational intelligence for biometrics and industrial applications.

Currently, he is an associate editor of the *IEEE Transactions on Human-Machine Systems* and the Springer Soft Computing. He served as associate editor of the *IEEE Transactions on Information Forensics and Security* and as guest coeditor for the *IEEE Transactions on Instrumentation and Measurement* (special section on *Computational Intelligence for Measurement Systems and Applications* and the special issue on *Biometric Instrumentation and Measurement*).

He was program chair for the IEEE Biometric Measurements and Systems for Security and Medical Applications (2011) and general co chair of the IEEE International Conference on Computational Intelligence for Measurement Systems and Applications (2011). He has served as program co chair for the IEEE Symposium on Computational Intelligence in Biometrics and Identity Management (2014), the IEEE Workshop on Biometric Measurements and Systems for Security and Medical Applications (2014, 2013, 2012), the IEEE International Conference on Computational Intelligence and Virtual Environments for Measurement Systems and Applications (2014, 2013), IEEE Workshop on Computational Intelligence in Biometrics and Identity Management (2011), the IEEE International Conference on Computational Intelligence for Measurement Systems and Applications (2012, 2010, 2008, 2006), and for the IEEE Workshop on Computational Intelligence in Biometrics: Theory, Algorithms, and Applications (2009). He has served on the program committees of various conferences. He is an IEEE senior member.

Chapter 1

Introduction

Biometrics is defined by the International Organization for Standardization (ISO) as "the automated recognition of individuals based on their behavioral and biological characteristics" [1]. Distinctive features evaluated by biometrics, which are referred to as biometric traits, include behavioral characteristics, such as signature, gait, voice, and keystroke, and biological characteristics, such as the fingerprint, face, iris, retina, hand geometry, palmprint, ear, and DNA.

Biometric recognition is the process by which the identity of a person is established and can be performed in two modalities: verification and identification. In the verification modality, the identity declared by an individual is compared with previously acquired biometric data. In contrast, in the identification modality, the recognition application determines a person's identity by comparing the acquired biometric data with information for a set of individuals.

Compared with traditional techniques for establishing identity, biometrics offers a greater confidence level that the authenticated individual is not impersonated by someone else. Traditional techniques are based on surrogate representations of identity, such as tokens, smart cards, and passwords, which can easily be stolen or copied, in contrast to biometric traits. This increased confidence enables the application of biometrics in a variety of scenarios, such as physical access control, government applications, forensic applications, and logical access control to data, networks, and services.

For real-world applications of biometric technologies, the acquisition process must be performed in a highly controlled and cooperative manner. To acquire good-quality biometric samples, users must perform deliberate actions, assume determinate poses, and remain motionless for a period of time. The applied scenario may also be subject to limitations, such as specific lighting and environmental conditions.

Biometric technologies that usually require constrained acquisitions include those based on characteristics of the face, iris, fingerprint, and hand. Traditional face recognition systems require users to assume a neutral pose and remain motionless for a period of time. The acquisition process uses a frontal camera under controlled light conditions. Iris acquisitions are typically performed at a distance of less than 30 cm from the camera and require that the users assume a defined pose and remain motionless while watching the camera. Moreover, these approaches use near-infrared illumination techniques, which can be perceived as a human health risk. Recognition systems based on fingerprint and hand characteristics require that the users touch the sensor surface with appropriate, uniform pressure. Contact with a sensor is often perceived as unhygienic and/or associated with a police procedure. This type of constrained acquisition technique can drastically reduce the usability and social acceptance of biometric technologies, thus decreasing the number of possible applicable contexts for biometric systems.

Other negative aspects of touch-based fingerprint recognition systems include the following: contact of the finger with the sensor platen introduces a security flaw due to the release of a latent fingerprint on the touched surface; the presence of dirt on the surface of the finger can reduce the accuracy of the recognition process; and different finger pressures on the acquisition area can introduce nonlinear distortions and low-contrast regions in the captured samples.

To increase the usability and social acceptance of biometric systems, researchers are studying less-constrained biometric recognition techniques based on different biometric traits, for example, face recognition systems in surveillance applications, iris recognition techniques based on images captured at a great distance and while in motion, and touchless technologies for fingerprint and hand characteristics.

Within this scenario, this book presents approaches for increasing the usability and social acceptance of fingerprint biometrics by performing less-constrained and highly accurate identity recognitions. In particular, approaches designed for high-security contexts are assessed to improve existing technologies adopted in border control, investigative, and governmental applications. Approaches based on low-cost hardware configurations are also analyzed with the aim of increasing the number of possible applicable scenarios for biometric systems.

The fingerprint is specifically considered in this book as a biometric trait because of its high distinctiveness and durability, dominant use in biometric applications, and wide range of applied contexts. The studied touchless biometric technologies are based on one or more cameras and use two-dimensional or three-dimensional samples. The primary goal is to perform accurate recognition in less-constrained application contexts compared to traditional fingerprint biometric systems. Other important goals include the use of a wider fingerprint area compared with traditional techniques, compatibility with the existing databases, usability, social acceptance, and scalability. The studied approaches are multidisciplinary because their design and realization have incorporated optical acquisition systems, multiple-view geometry, image processing, pattern recognition, computational intelligence, and statistics.

The considered biometric technologies have been applied to different biometric datasets that represent a heterogeneous set of application scenarios, confirming the feasibility of the studied approaches. Compared with traditional fingerprint recognition systems, the realized touchless biometric systems offer better accuracy, usability, user acceptability, scalability, and security.

1.1 State of the Art

This book presents innovative, less-constrained biometric technologies that perform identity recognition with high accuracy. Some of these techniques are particularly suitable for consumer applications.

Studies of touchless fingerprint recognition techniques are limited, and reported techniques can be divided into two classes: methods based on two-dimensional samples and methods based on three-dimensional samples. The former methods usually capture single fingerprint images using single CCD cameras. However, systems based on three-dimensional models require more complex hardware setups but can provide greater accuracy.

Most reported touchless fingerprint recognition systems comprise three primary steps: acquisition, computation of a touch-equivalent fingerprint image, and feature extraction and matching. Different touchless acquisition strategies based on single CCD cameras, multiple-view techniques, or structured-light approaches exist. In systems based on two-dimensional samples, the computation of touch-equivalent fingerprint images is usually performed by applying enhancement algorithms and normalizing resolution. In contrast, most systems based on three-dimensional samples perform an unwrapping step that aims to map the three-dimensional fingerprint models into two-dimensional space. Then, the obtained touch-equivalent images are processed using traditional feature extraction and matching methods designed for touch-based fingerprint images.

The majority of reported touchless fingerprint recognition systems are based on complex and expansive hardware setups, use finger placement guides, and usually perform acquisitions at distances to the cameras of less than 10 cm.

1.2 The Performed Research

The realized approaches include techniques for all steps of touchless fingerprint recognition systems based on two-dimensional and three-dimensional samples. The first class of techniques is based on single fingerprint images and is designed to be adopted using mobile devices with integrated cameras. In contrast, techniques based on three-dimensional samples require more complex acquisition setups involving two cameras and offer more accurate results.

Studied techniques for touchless biometric systems based on two-dimensional samples include acquisition setups that do not require the use of finger placement guides and can capture biometric samples at a greater distance compared with other techniques in the literature; a quality assessment method for touchless fingerprint images to select the best-quality frames in the obtained frame sequences representing less-constrained fingerprint acquisitions; techniques for the global analysis of touchless images; methods that permit touch-equivalent fingerprint images compatible with traditional feature extraction and matching algorithms to compute touch-based acquisitions; and techniques for reducing the effects of perspective distortions and three-dimensional rotations of the finger with respect to the acquisition camera.

The implemented fingerprint recognition techniques based on three-dimensional samples require different acquisition setups based on multiple-view methods, three-dimensional reconstruction approaches, feature extraction, and matching algorithms. For identity recognition based on three-dimensional samples, two approaches are available: the first extracts and compares features related to the three-dimensional coordinates of minutiae points, and the second computes touch-equivalent images and then applies biometric recognition algorithms designed for touch-based images to obtain fingerprint images compatible with existing biometric databases.

A technique for the computation of synthetic touchless samples has been studied to reduce the efforts necessary to collect the biometric data required to design and test new algorithms and hardware setups.

1.3 Results

The implemented methods have been analyzed using datasets captured under different application conditions, confirming the feasibility of the proposed approaches within the considered application contexts. In particular, touchless recognition systems based on two-dimensional samples have obtained satisfactory performance in low-cost applications, whereas touchless systems based on three-dimensional samples have achieved good accuracy in high-security applications.

The implemented fingerprint recognition approaches have also been compared with traditional touch-based systems by performing a multidisciplinary test that considers the following set of important aspects of biometric systems: accuracy, speed, cost, scalability, interoperability, usability, social acceptance, security, and privacy. These comparisons have demonstrated that in high-security applications, the realized approaches based on three-dimensional samples are more accurate than touch-based systems. In addition, touchless recognition techniques enhance scalability, usability, social acceptance, and security compared with traditional fingerprint recognition systems.

1.4 Structure of This Book

This book is structured as follows:

- Chapter 2 presents a brief description of the following aspects of biometric systems: biometric traits, applications, evaluation techniques, and research trends.
- Chapter 3 discusses recent studies of less-constrained biometric systems based on traits including the face, iris, gait, and ear; soft biometric traits; and hand characteristics.
- Chapter 4 provides a literature review of fingerprint recognition systems. The steps of the biometric recognition process in well-known techniques in the literature are analyzed, followed by a discussion of recent studies regarding touchless recognition systems based on two-dimensional and three-dimensional samples.
- Chapter 5 describes research regarding touchless fingerprint recognition systems, including the details of methods designed for two-dimensional samples, approaches based on three-dimensional fingerprint samples, and computational techniques for synthetic touchless samples.
- Chapter 6 describes experiments involving the studied touchless fingerprint recognition approaches. In particular, this chapter analyzes the performance of every implemented technique and presents a comparison of the realized touchless biometric systems with touch-based fingerprint recognition technologies.
- Finally, Chapter 7 presents the conclusions and future work.

Chapter 2

Biometric Systems

Traditional techniques used to establish the identity of a person are based on surrogate representations of his/her identity, such as passwords, keys, tokens, and identity cards. In contrast, biometric recognition systems are based on physiological or behavioral characteristics of the individual that are unequivocally related to the owner, cannot be shared or misplaced, and are difficult to steal. Systems based on physiological traits perform and analyze measurements of a part of the human body, whereas those based on behavioral traits evaluate actions.

The most suitable biometric recognition technology for an application scenario must be selected by evaluating different factors, as discussed in the literature. Recent research in biometrics has aimed to improve the accuracy, reduce the hardware costs, and increase the usability and user acceptability of existing technologies.

Section 2.1 discusses the general characteristics of biometric systems and traits. Then, the application contexts of biometric technologies are discussed in Section 2.2, and the most used techniques in the literature for evaluating biometric applications are described in Section 2.3. Finally, Section 2.4 briefly presents the most important research trends.

2.1 Biometric Traits

Biometric recognition or, simply, biometrics, refers to the automatic recognition of individuals based on their physiological and/or behavioral characteristics. Biometrics allow an individual's identity to be confirmed or established based on "who she is" rather than by "what she possesses" (e.g., an ID card) or "what she remembers" (e.g., a password) [2]. Examples of physiological biometric traits include the fingerprint, hand geometry, iris, retina, and face. The distinctiveness of a signature or gait can be considered behavioral biometric traits. Figure 2.1 presents some examples of biometric traits.

A physiological or behavioral characteristic can be considered a biometric trait only if this characteristic satisfies the following four conditions [2]:

1. *Universality*: Every person should possess the characteristic.
2. *Distinctiveness*: Two persons should not have the same characteristic, or the probability of this event should be negligible. The differences between characteristics of different individuals should also be sufficient to discriminate their identities.
3. *Permanence*: The characteristic should be sufficiently invariant over a period of time. The considered period of time depends on the application context.
4. *Collectability*: The characteristic should be quantitatively measurable.

The following three important characteristics of biometric traits are more related to the technologies adopted to perform the identity recognition and should be high:

1. *Performance*: The accuracy, speed, and robustness of the considered technology.
2. *Acceptability*: The acceptance level of the biometric technology by the user.
3. *Circumvention*: The robustness of the technology against fraudulent techniques.

In the literature, some of the most commonly used physiological biometric traits include the following: fingerprint [5,6], iris [7,8], face [9], hand geometry [10], palmprint [11], palm vein [12], and ear [13,14].

Some of the most commonly used behavioral traits are the following: voice [15,16], signature [17], gait [18], and keystroke [19].

Recently, soft biometric traits have also been studied to increase the accuracy of biometric recognition systems or to perform uncooperative recognitions. Soft biometric traits are characteristics that provide information about the individual, but lack the distinctiveness and permanence to sufficiently differentiate any two individuals [20]. Soft biometric traits can be continuous or discrete. Continuous traits are characteristics of the human body measured in a continuous scale, whereas discrete traits represent discrete classifications of distinctive characteristics. Examples of continuous characteristics include the following: height [21], weight [22], and other measurements of the body parts [23]. Examples of discrete soft biometrics include the following: gender [24,25], race [24], eye color [26], and clothing color [23].

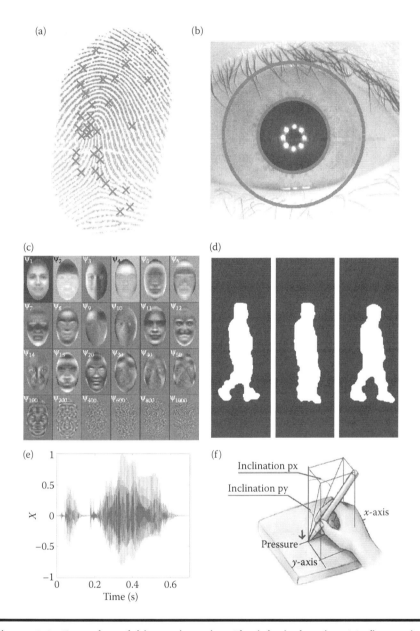

Figure 2.1 Examples of biometric traits. Physiological traits: (a) fingerprint, (b) iris, and (c) face. (M. Meytlis and L. Sirovich, On the dimensionality of face space, *IEEE Transactions on Pattern Analysis and Machine Intelligence*, vol. 29, no. 7, pp. 1262–1267. © 2007, IEEE.) Behavioral traits: (d) gait, (e) voice, and (f) signature. (D. Muramatsu et al., A Markov chain Monte Carlo algorithm for Bayesian dynamic signature verification, *IEEE Transactions on Information Forensics and Security*, vol. 1, no. 1, pp. 22–34. © 2006, IEEE.)

2.2 Applications

In recognition applications, a biometric system can be considered a pattern recognition system that performs personal recognition by determining the authenticity of a specific physiological or behavioral characteristic possessed by the user.

Biometric systems can function in two different modalities: verification and authentication. Verification involves confirming or denying a person's claimed identity. The system performs a one-to-one comparison of the acquired biometric data with the stored information associated with the claimed identity. An individual who wants to be recognized usually claims an identity via a personal identification number (PIN), a user name, or a smart card.

In identification mode, the biometric system must establish a person's identity by performing a one-to-many comparison of the acquired biometric data with the information for a set of individuals. The identification does not require the user to claim an identity. Two classes of identification exist: positive identification and negative identification. The first class addresses the question "Who am I?" Typically, the returned results are numerical identifiers or access permissions. The second class searches databases in the same fashion, comparing one template against many, to ensure that a person is not present in a database. These systems are usually adopted for security controls in public buildings (e.g., stadiums, swimming pools, and train stations). The term "recognition" is generic and does not make a distinction between verification and identification.

Both the verification and identification modalities require a previous step in which user biometric data are stored in a biometric document or database. This step is called enrollment and usually requires the presence of a certified authority. The schemas of the verification, identification, and enrollment modalities are presented in Figure 2.2. These processes are based on the following common components: a biometric sensor, feature extractor, database, matcher, and decision module.

- The acquisition sensor used is dependent on the biometric trait. The acquired data are usually images, signals, or frame sequences. These data are called biometric samples.
- The feature extraction module aims to extract a template, which is an abstract and distinctive representation of the biometric sample. The computed characteristics can differ among systems based on the same biometric trait.
- The database contains the templates of the enrolled individuals. Many systems use centralized databases to perform verifications or identifications. Different storage supports, such as smart cards or USB devices, can also be used in verification systems.
- The matcher compares two or more biometric templates, obtaining a value called a matching score. The template comparison can be based on different metrics and can return a similarity or dissimilarity index.

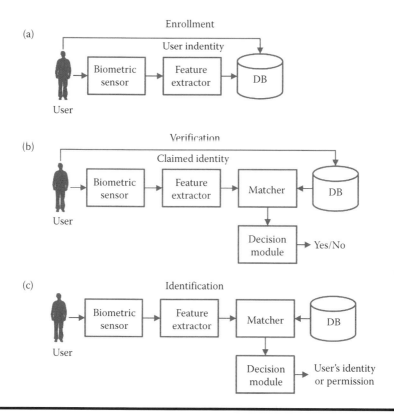

Figure 2.2 Modalities of biometric systems: (a) enrollment, (b) verification, and (c) identification.

■ The decision module uses the computed matching score to determine whether a template appertains to an individual. The decision is obtained by applying a threshold to the matching score. The threshold value must be accurately tuned for the application context because this value determines the tradeoff between false positives and false negatives.

The use of biometric systems in different application scenarios is continuously increasing [27]. A significant positive trend is apparent in the related market. In fact, this market reached 5 billion dollars in 2011 and is expected to reach 12 billion dollars by the end of 2015 [28].

Typical application scenarios include physical access control (in critical areas, public buildings, sport arenas, bank vaults, transportation systems, etc.), surveillance (of private buildings, public areas, etc.), government applications (identity cards, passports, driving licenses, immigration control, health cards, access control to online government services, etc.), forensic applications (body identification, crime investigation, missing person searches, kinship, intelligence, etc.), and logical access control to data, networks, and services (home banking, ATM, supermarkets, e-commerce, mobile phones, computers, etc.).

The most commonly used biometric trait is the fingerprint [27]. In fact, this trait was the first human characteristic used to perform biometric recognitions. Fingerprint recognition systems are also characterized by good performance in terms of accuracy and speed. Some automated fingerprint identification systems (AFIS) can recognize millions of templates [29]. Most fingerprint recognition systems evaluate the position and shape of local ridge patterns (minutiae); however, some systems use other features and correlation-based techniques [5].

Other widely used biometric systems are based on facial characteristics. These systems are less accurate than fingerprint recognition techniques. However, these systems are characterized by greater user acceptance because humans usually perform recognition by evaluating facial characteristics. Moreover, touchless acquisition systems are considered less invasive. Face recognition techniques can also be used in different application contexts than fingerprint recognition systems, such as surveillance and entertainment applications. Biometric systems based on facial characteristics can use transformation-based techniques or attribute-based methods [9,30].

Iris recognition systems are considered the most rapid and accurate biometric techniques. Thus, these systems have been widely applied in border control and airports. The primary problem with these systems is the use of constrained acquisition techniques, which can be considered invasive. Thus, these systems are characterized by low user acceptance. To overcome this limitation, researchers are studying systems that can function under less collaborative conditions [31,32]. One of the most used recognition methods is based on a feature representation called Iriscode [8].

Additionally, some diffused biometric systems are based on hand characteristics, such as hand geometry [10] or the palmprint and palm vein [11,12]. Systems based on hand geometry are usually adopted in application contexts in which a high security level is not necessary. These systems do not have high accuracy but are characterized by high user acceptance and low hardware costs. Systems based on palmprint and palm vein characteristics can obtain results that are more accurate.

Some promising systems based on other physical traits use DNA [33] and ear shape [13,34]. DNA is characterized by high accuracy; however, the recognition techniques based on this trait are expensive and require long evaluation times. Biometric systems based on ear shape are being studied because these systems can obtain sufficiently accurate results using touchless acquisitions performed at long distances.

Diffused systems based on behavioral traits consider the characteristics of the voice [15,16], signature [17], keystroke [19], and gait [18]. These systems are characterized by high user acceptability, but are usually less accurate than systems based on physiological characteristics.

Multimodal and multibiometric systems, which fuse information obtained regarding different traits and/or recognition algorithms to increase the obtained

Table 2.1   Properties of Biometric Systems Related to the Used Biometric Trait

Trait	Univ.	Uniq.	Perm.	Coll.	Perf.	Acc.	Circ.
Face	H	L	M	H	L	H	L
Fingerprint	M	H	H	M	H	M	H
Hand geometry	M	M	M	H	M	M	M
Keystrokes	L	L	L	M	L	M	M
Palm vein	M	M	M	M	M	M	H
Iris	H	H	H	M	H	L	H
Retinal scan	H	H	M	L	H	L	H
Signature	L	L	L	H	L	H	L
Voice	M	L	L	M	L	H	L
Face thermograms	H	H	L	H	M	H	H
Odor	H	H	H	L	L	M	L
DNA	H	H	H	L	H	L	L
Gate	M	L	L	H	L	H	M
Ear	M	M	H	M	M	H	M

Source: A. Jain, A. Ross, and S. Prabhakar, An introduction to biometric recognition, *IEEE Transaction on Circuits and Systems for Video Technology,* vol. 14, no. 1, pp. 4–20. © 2004, IEEE.

Notes: Univ.: universality; Uniq.: uniqueness; Perm.: permanence; Coll.: collectability; Perf.: performance; Acc.: acceptability; Circ.: circumvention; H: high; M: medium; L: low.

recognition accuracy, have also been discussed [35–39]. Such systems are usually adopted in high-security application contexts.

Other recognition applications use soft biometrics. Soft biometric traits are physical, behavioral, or adhered human characteristics that usually do not permit unequivocal recognition [20]. However, these traits can be used to perform unobtrusive identifications among a limited number of users, as a preliminary screening filter, or in combination to increase the recognition accuracy of biometric systems.

Table 2.1 reports the principal characteristics of biometric systems based on the most commonly used biometric traits in the literature.

2.3 Evaluation of Biometric Systems

The evaluation of biometric systems is a complex multidisciplinary task. The evaluation aspects depend on the application context. Accuracy, speed, cost, usability, acceptability, scalability, interoperability, security, and privacy must also be considered. Accuracy is usually one of the most important aspects to be considered during the design of biometric applications and should be evaluated using specific techniques and figures of merit.

2.3.1 Evaluation Strategies

Biometric applications should be analyzed using different evaluation strategies to consider all characteristics of the technologies used. The evaluation strategies can be divided into the following three overlapping categories, with respect to increasing complexity of uncontrolled variables: technology, scenario, and operational [40].

Technology evaluation aims to measure the performance of biometric techniques. Typically, this category of evaluations is used to compare the performance of biometric algorithms on reference datasets. Publically available datasets are usually adopted to compare the performance obtained with that achieved by other reported algorithms. An example of a technology evaluation is the Fingerprint Verification Competition [41].

Scenario evaluation considers additional variables related to the application context. The goal is to measure the performance of different biometric systems in a particular application. An example of such an evaluation is the comparison of the performances obtained by biometric systems based on different traits for access control in a laboratory. Typically, every compared system captures the samples with different sensors. Using data acquired from the same individual is preferable. The creation of the testing datasets is usually a long process that can require several weeks. In fact, this task may require a statistically significant number of samples to be obtained and may be based on different trials to ensure adequate user habituation. Considering these factors, scenario evaluations are not always repeatable. The UK Biometric Product Testing is an example of a large-scale scenario evaluation [42].

Operational evaluation differs from scenario evaluation because this evaluation is performed in a real application scenario. Therefore, operational evaluation can be performed using all of the effective users of the system, a randomly selected set of users, or specifically selected users. Thus, the performed evaluation is difficult or impossible to repeat. The goal of this type of evaluation is not to measure the accuracy of the biometric system, but to determine the impact on workflow of the addition of a biometric system in the considered application context. Operational evaluation can thus facilitate the analysis of the advantages of the considered biometric system.

The described evaluation modalities are complementary and should be performed in sequence. First, technology evaluation analyzes biometric method performance under general conditions. Then, scenario evaluation permits the selection of the best biometric technology for an application scenario. Finally, operational evaluation yields solid business reports for potential installations.

2.3.2 Evaluation Aspects

Biometric systems should be evaluated from different perspectives. As presented in Figure 2.3, the primary evaluation aspects can be quantified and plotted in a

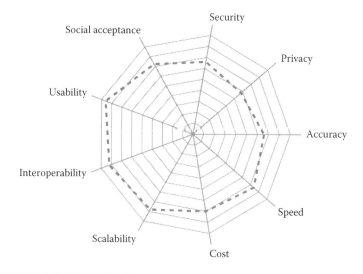

Figure 2.3   Evaluation aspects of biometric systems.

nine-dimensional space (e.g., in a spider diagram), in which a specific system can be represented by a point.

- *Accuracy*: Specific figures of merit should be adopted to measure the performance of the recognition system. The evaluation procedures are discussed in Section 2.3.3.
- *Speed*: The time required for each recognition is an important aspect to consider during the design of a biometric application. For example, faster algorithms are needed for identification systems than for verification systems. All steps of the biometric recognition process should be considered: acquisition, feature extraction, and matching.
- *Cost*: Systems based on expensive, accurate acquisition devices usually feature better performance than systems based on low-cost hardware setups. However, low-cost sensors can permit a wider diffusion of biometric technology. To identify the best biometric technique for the considered application context, a cost analysis should be performed that includes the costs of the design and software implementation.
- *Scalability*: This aspect refers to the ability of the system to work under an increased or expanding workload. In biometrics, this aspect should be evaluated by considering all modules of the system: the network architecture, sensor availability, storing device type(s), and recognition algorithm performance in terms of accuracy and speed.
- *Interoperability*: This aspect represents the level of compatibility between the evaluated system and existing biometric technologies. The use of standard formats for templates and samples permits compatibility between different systems based on the same trait [43].

- *Usability*: This aspect refers to the ease of use and learnability of the system. Low usability levels can reduce social acceptance and system accuracy. Usability is typically evaluated by analyzing the acquisition time, number of incorrectly captured samples, and questionnaire results [44].
- *Social acceptance*: This aspect measures users' perceptions, feelings, and opinions regarding the system. This factor is crucial for the diffusion of a biometric technology in different application contexts. The acceptance level may be different for different users and is influenced by various factors, such as cultural and religious aspects, usability, and perceived privacy risks. Social acceptance can be measured by analyzing the results of questionnaires [45].
- *Security*: This aspect measures the robustness of the system against possible attacks. Security ensures verification, data integrity, confidentiality, and non-repudiation. All modules of biometric systems are susceptible to attack [46,47]. Intrinsic aspects of the adopted technologies and the use of countermeasures to prevent possible attacks influence the security level of the system. An important aspect is the robustness of the system against spoofing attacks [48]. These aspects become particularly relevant in many emerging distributed infrastructures and applications (e.g., in cloud computing environments [49–54]).
- *Privacy*: This aspect measures the capability of the system to prevent possible thefts and misuses of biometric data [55,56]. In contrast to security, privacy also requires data protection [57–60]. Evaluating the privacy protection level is a complex task and should consider both real risks and the risks perceived by the users.

2.3.3 Accuracy Evaluation

The presented accuracy evaluation methodologies are based on the analysis of the results obtained by recognition systems using datasets of biometric samples and can be applied to technologies based on symmetrical and asymmetrical matching functions.

The obtained results permit the calculation of the most common figures of merit discussed in the literature. These accuracy indices can then be used to compare the performances of existing biometric recognition techniques.

However, the accuracy indices express the performances of the evaluated biometric systems using limited sets of samples. Thus, estimating the confidence of the obtained measurements is good practice.

2.3.3.1 Methods for Accuracy Evaluation

The reported procedures and figures of merit can be used to perform the technology evaluations and scenario evaluations described in Section 2.3.1.

Accuracy evaluation requires the use of a previously captured biometric database. Technology evaluations are usually performed using publicly available datasets to compare the obtained results with those achieved by other methods in the literature. Scenario evaluations typically do not use public datasets, but require sets of specifically acquired biometric data to describe the evaluated scenario in an exhaustive manner.

Studies have examined the accuracy evaluation of symmetric matching functions [61]. For two biometric samples A and B, a matching function M is symmetric if $M(A, B) = M(B, A)$. The evaluation procedures reported here are those proposed in [61] and can be applied to both symmetric and asymmetric matching functions, in which $M(A, B) \neq M(B, A)$. Accuracy evaluation techniques that consider asymmetric matching functions are necessary for fingerprint biometrics because most recognition systems based on minutiae features use asymmetric matching strategies.

We define B_{ij} as the jth sample of the ith individual, T_{ij} as the biometric template obtained from B_{ij}, n_i as the number of samples pertaining to the individual i, and N as the number of identities enrolled in the database. Then, we describe the steps required to perform the accuracy evaluation using the most common figures of merit described in the literature.

1. *Enrollment*

    The enrollment phase computes and stores the templates $T_{ij}, i = 1, \ldots, N$, $j = 1, \ldots, n_i$. The presence of errors in this phase is described using the index $\text{REJ}_{\text{ENROLL}}$, which represents the rejection ratio due to fails (software modules deny the enrollment of the acquired sample) and crashes (software or hardware).

2. *Recognition*

    The procedure to evaluate symmetric matching functions is as follows: identity comparisons are performed between every template T_{ij} pertaining to the biometric dataset and the templates $T_{ik}(j < k \leq n_i)$, obtaining a matrix called genuine matching scores gms_{ijk}. The term "genuine" defines identity comparisons performed between templates for the same individual i. Only the upper triangular matrix of the squared matrix gms is computed because gms is symmetric by definition. This procedure computes a gms matrix for each individual.

    In contrast, for asymmetric matching functions, the genuine matching scores matrix gms_{ijk} is computed by performing identity comparisons between every template T_{ij} pertaining to the biometric dataset and the templates $T_{ik}(j \leq n_i, k \neq i)$. The obtained matrix remains square but is not symmetric.

    For symmetric matches, the number of genuine recognition attempts (NGRA) is defined as

$$\text{NGRA}_{\text{symMatch}} = \frac{1}{2} \sum_{i=1}^{N} n_i(n_i - 1), \tag{2.1}$$

    where $\text{REJ}_{\text{ENROLL}} = 0$.

In contrast, NGRA for asymmetric matches is defined as

$$\text{NGRA}_{\text{asymMatch}} = \sum_{i=1}^{N} n_i(n_i - 1), \qquad (2.2)$$

where $\text{REJ}_{\text{ENROLL}} = 0$.

Accuracy evaluation also considers identity comparisons between templates related to different individuals (impostor matching).

For symmetric matching functions, identity comparisons are performed between the templates $T_{i1}, i = 1, \dots, N$, and the first biometric template of different individuals $T_{k1}(1 < k \leq N, k > i)$, to obtain a matrix called impostor matching scores ims_{ijk}.

For asymmetric matching functions, identity comparisons are performed between the template $T_{i1}, i = 1, \dots, N$, and the first biometric template of different individuals $T_{k1}(1 < k \leq N, k \neq i)$, obtaining the matrix ims_{ijk}.

For symmetric matching functions, the number of impostor recognition attempts (NIRA) is defined as

$$\text{NIRA}_{\text{symMatch}} = \frac{1}{2} N(N - 1), \qquad (2.3)$$

where $\text{REJ}_{\text{ENROLL}} = 0$.

In contrast, NIRA for asymmetric matching functions is defined as

$$\text{NIRA}_{\text{asymMatch}} = N(N - 1), \qquad (2.4)$$

where $\text{REJ}_{\text{ENROLL}} = 0$ for asymmetric matching.

The described procedure does not consider errors during the enrollment phase. Nevertheless, this information can be used by accumulating the REJ_{NGRA} and REJ_{NIRA} counters into the gms and ims matrices, respectively. Then, the matrices gms and ims can present missing values, which are commonly stored as "NULL" or as numbers out of range with respect to the scores that can be returned by the matching function.

2.3.3.2 Accuracy Indexes

The most common figures of merit for the accuracy evaluation of biometric systems are briefly discussed in this section.

Some biometric systems permit multiple attempts or use multiple enrolled templates. For simplicity, the reported general definitions refer to the common case in which the system permits a single attempt and uses a single enrolled template for each individual.

Commonly used metrics are the false match rate $\text{FMR}(t)$ and false non-match rate $\text{FNMR}(t)$, which are functions of the threshold value t used in the decision step of the biometric recognition process. FMR is the expected probability that an impostor is erroneously considered a genuine match. FNMR is the expected probability that a genuine match is erroneously considered an impostor.

The FMR is the expected probability that a sample will be falsely declared to match a single randomly selected template (false positive). The FNMR is the expected probability that a sample will be falsely declared not to match a template of the same measure from the same user supplying the sample (false negative).

The functions FMR and FNMR are computed from the gms and ims matrices by varying the decision threshold t and are defined as follows [5]:

$$FMR(t) = \frac{card(\{ims_{ik} | ims_{ik} \geq t\})}{NIRA}, \quad (2.5)$$

$$FMR(t) = \frac{card(\{gms_{ijk} | gms_{ijk} < t\}) + REJ_{NGRA}}{NGRA}, \quad (2.6)$$

where card($\cdot$) is the cardinality.

The overall accuracy level of a biometric system is often evaluated by considering two error plots. The first plot is the receiving operating curve (ROC), which is a graphical plot of the fraction of true positives versus the fraction of false positives for a binary classification system with a varying discrimination threshold. In biometrics, the axes are 1-FNMR and FMR. The second graph is the plot of FNMR versus FMR, which is called the detection error tradeoff (DET) plot. Many studies use logarithmic axes for these graphs.

The ROC and DET curves can be used to directly compare biometric systems. To analyze the accuracy of different systems, the errors must be evaluated using the same dataset. Figure 2.4 presents the DET curves that compare the accuracy of four different systems. The DET curve of the best system is below all other curves. Typically, a system outperforms compared systems only in some regions determined by specific values of t.

Another frequently used graph plots the distributions of the matching scores obtained using genuine (gms_{ijk}) and impostor (ims_{ik}) identity comparisons. If the two curves do not intersect in the graph, then the biometric system properly recognizes all samples pertaining to the evaluated dataset, presenting a threshold value t' that perfectly classifies the genuine and impostor matching scores. Figure 2.5 presents an example of a graph obtained by an iris recognition system [63] using a public dataset [64].

Another commonly used accuracy index is the equal error rate (EER), which is the ideal point at which $FMR(t) = FNMR(t)$.

Other figures of merit describe the accuracy of the acquisition and enrollment processes: the failure to accept rate (FTAR) is the expected proportion of transactions for which the system is unable to capture or locate an image or signal of sufficient quality [5]; the failure to enroll rate (FTER) represents the expected proportion of the population for whom the system is unable to generate repeatable templates [5].

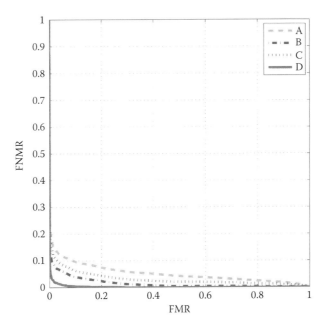

Figure 2.4 Examples of DET curves.

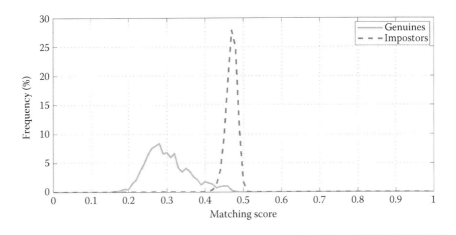

Figure 2.5 Examples of genuine and impostor distributions.

Many other indices described in the literature are useful for testing the performance of biometric technologies. The choice of indices depends on the system characteristics (identification/verification, the fixed threshold, the number of enrolled users, and the number of templates per user) [5]. Common figures of merit are the false acceptance rate (FAR) and false reject rate (FRR). The FAR represents the frequency with which a nonauthorized person (system user) is accepted as authorized. The FRR is the frequency with which an authorized person is erroneously rejected. In contrast to the FMR and FNMR, the FAR and FRR are computed based on the number of incorrect acquisitions.

2.3.4 Confidence Estimation

The accuracy evaluation of a biometric system is performed using limited sets of data. Thus, estimating the confidence of the performed measures is useful.

The confidence of the accuracy indices is strictly related to the testing dataset. As reported in [65], the datasets used to perform the evaluation should have a sufficient number of samples. In choosing the size of the test dataset, two primary rules should be considered: the Rule of 3 [66,67] and Rule of 30 [68]. The Rule of 3 establishes the lowest error rate that can be statistically obtained for a given number N of identity comparisons to be estimated. This error rate corresponds to the value p for which the probability of zero errors in N trials is equal to a fixed value (typically 5%). This rule can also be expressed as $p \approx 3/N$ for a 95% confidence level. For example, a test without errors on a dataset of 300 independent samples suggests with 95% confidence that the error rate of the evaluated biometric system is equal to or less than 1%.

The Rule of 30 states that a test must obtain at least 30 errors to be 90% confident that the true error rate is within ±30% of the obtained error rate. For example, 30 false non-match errors in 3000 independent genuine trials suggest with 90% confidence that the true error rate is between 0.7% and 1.3%. This rule is derived from a binomial distribution assuming independent trials. The central limit theorem [69], which approximates the obtained error rates to normal distributions, can be applied to sufficiently large sample datasets. Under the assumption of normality, $100(1 - \alpha)$% confidence bounds on the obtained error rates can be computed as

$$\hat{p} \pm z\left(1 - \frac{\alpha}{2}\right)\sqrt{\hat{V}(\hat{p})}, \tag{2.7}$$

where $\hat{p}$ is the obtained error rate, $\hat{V}(\hat{p})$ is the variance of the obtained error rate [65], and $z(\cdot)$ is the inverse of the standard normal cumulative distribution. For 95% confidence limits, the value $z(0.975)$ is 1.96.

Nonparametric methods such as the bootstrap method [70,71] have also been described. This technique reduces the requirement to make assumptions regarding the underlying distribution of the observed error rates and the dependencies between attempts. The distributions and dependencies are inferred from the samples themselves. Bootstrap samples are computed by sampling with replacement from the original samples. Then, many bootstrap samples are used to obtain the empirical distribution of the accuracy indices, which can then be used to estimate the confidence intervals. The bootstrap values permit the direct construction of $100(1 - \alpha)$% confidence limits by choosing L (lower limit) and U (upper limit) such that only a fraction $\alpha/2$ of bootstrap values are lower than L and that $\alpha/2$ bootstrap values are higher than U. The work in [65] recommends using at least 1000 bootstrap samples for 95% limits and at least 5000 bootstrap samples for 99% limits.

The approach proposed in [72] is based on the bootstrap technique, but also considers that the templates related to the same biometric traits and individuals are statistically correlated. This approach samples with replacement from specifically determined subsets of the data. Every subset contains the templates obtained from a single biometric trait or individual.

Other works [73,74] evaluate the robustness of bootstrap strategies under different application conditions.

Studies have also examined the confidence estimation of biometric systems based on different techniques. A semiparametric approach based on multivariate copula models for correlated biometric acquisitions was proposed in [75]. Based on the same model, this work also proposed a technique to determine the minimum number of samples required to achieve confidence bands of a desired width for the ROC curve. Other studies have analyzed multibiometric systems [76].

2.4 Research Trends

To resolve the limitations of existing biometric systems, the international research community is extremely active in the following major research fields:

- *Increased accuracy*: Many studies have aimed at improving the accuracy of biometric systems using new software and hardware techniques.
- *Multimodal and multibiometric systems*: These systems use multiple biometric traits and/or multiple recognition algorithms to obtain greater accuracy compared to traditional biometric systems [35,39,77]. The information obtained from every biometric technology is fused using decisional techniques (e.g., computational intelligence [36]).
- *Reduced sensor costs*: Many existing biometric systems are based on expensive sensors. For example, most iris recognition systems capture iris images using near-infrared illuminators and cameras [8]. Systems based on images captured under natural light conditions can reduce final costs. In this context, for example, studies have examined segmentation [32,78] and recognition [79] of iris images captured under natural light conditions.
- *Less-cooperative acquisition techniques*: Existing biometric methods usually require a high level of cooperation during the acquisition step. Technologies based on less-cooperative acquisition techniques permit the adoption of biometric systems under different application scenarios, such as surveillance applications [80].

■ *Increased usability and user acceptance*: More usable systems and greater user acceptance can permit wider diffusion of biometric systems. Studies have examined the usability [44] and the user acceptance [45] of biometric systems. The use of less-cooperative touchless acquisition techniques can improve these aspects of biometric systems.

■ *Increased distance from sensors*: Biometric techniques based on samples captured at a greater distance compared with traditional systems can permit the reduction of acquisition time and the development of new biometric applications. As an example, the system described in [81] is based on iris images captured at a distance of 3 m between the user and the sensor and is designed to work in biometric portals.

■ *Three-dimensional sample acquisition*: Three-dimensional models of biometric traits do not exhibit distortions associated with the mapping of three-dimensional objects into a two-dimensional space. Moreover, these models feature higher accuracy compared to traditional biometric techniques because additional information related to the z-axis can be used during the feature extraction and matching steps. Most studies of the use of three-dimensional samples have focused on facial traits [82].

■ *Security and privacy*: Protecting all of the components of biometric systems and user data are important factors for wider diffusion of biometric technologies. Existing techniques can reduce the performance in terms of accuracy and speed. Studies have focused on overcoming this limitation [83–86].

■ *New biometric traits*: Novel traits should permit the use of biometrics in new applications. For example, recent studies have analyzed the use of cardiac signals [87–89].

The primary focus of this book is less-constrained biometric systems and techniques for privacy protection. The term "less-constrained" encompasses techniques that increase usability and user acceptance, increase distances from sensors, and reduce the level of cooperation required during biometric acquisitions.

2.5 Summary

Traditional techniques for recognizing individuals are based on surrogate representations of identities, such as passwords, keys, tokens, and identity cards. These representations cannot guarantee a sufficient level of security for some applications because of the possibility of sharing, loss, or theft. In contrast, biometric recognition systems are based on physiological or behavioral characteristics of the individual that are univocally related to their owner, cannot be shared or misplaced, and are more difficult to steal. Examples of physiological biometric traits include the face, fingerprint, iris, hand geometry, palmprint, and ear. Examples of behavioral characteristics include the voice, signature, gait, and keystroke. Studies have also examined recognition techniques based on less distinctive characteristics, referred to as soft biometric traits. Recognition techniques based on soft biometric characteristics can be applied to increase the accuracy of existing biometric systems or to perform recognition under covert and uncooperative situations. Examples of soft biometric traits include height, weight, gender, and race.

Biometric systems can perform identity recognition in two different modalities: verification and identification. In verification mode, the acquired biometric data are compared with stored information associated with the claimed identity. In identification mode, a person's identity is established by comparing the acquired biometric data with the information related to a set of individuals. Both verification and identification systems are based on the following common components: a biometric sensor, feature extractor, database, matcher, and decision module.

The use of biometric systems is continuously increasing under different application scenarios. Typical application scenarios include physical access control, government applications, and forensic applications, as well as logical access control to data, networks, and services. The most commonly used biometric trait is the fingerprint because fingerprint recognition systems are characterized by low cost and good performance in terms of accuracy and speed. Face recognition systems have also achieved widespread use because these systems have greater user acceptance. Iris recognition systems are considered to be the fastest and most accurate biometric techniques, but rely on complex acquisition procedures. Thus, these systems are principally used in high-security applications, such as border control and airports. Other diffused biometric technologies are based on hand characteristics or behavioral traits. The choice of the most suitable biometric technology is strictly dependent on the application context and should be based on accurate analyses. In fact, the evaluation of a biometric system is a complex and multidisciplinary task that should be performed considering different perspectives related to the recognition technique and application scenario. Important aspects that should be considered include accuracy, speed, cost, scalability, interoperability, social acceptance, usability, security, and privacy. System accuracy is usually one of the most important aspects that should be considered during the design of a biometric application and should be evaluated using specific techniques and figures of merit. However, the obtained results are related to limited sets of data. Thus, estimating the confidence of the obtained accuracy indices is a common practice.

The international research community is extremely active in studying methods to improve current biometric systems. Some important research trends include increasing the accuracy of existing techniques, studying new biometric traits, designing multimodal and multibiometric systems, reducing hardware costs, increasing the usability and acceptability of biometric technologies, increasing the distance between the user and sensor(s), providing security and privacy protection, designing less-constrained acquisition techniques, and implementing recognition techniques based on three-dimensional samples.

Chapter 3

Touchless and Less-Constrained Biometrics

Most biometric applications are based on highly controlled, constrained acquisition procedures. For example, biometric systems can require that the users touch a sensor and remain motionless for a period of time.

Researchers are studying techniques to reduce the constraints imposed by biometric acquisition procedures. These techniques may permit the extension of biometrics to new application contexts and increase the usability and user acceptance of recognition applications.

Many studies have examined less-constrained biometric systems based on biometric characteristics traditionally captured by CCD cameras (e.g., the face, iris, gait, and ear, as well as soft biometric traits). The goal of these studies has been to design biometric systems compatible with covert applications, which may involve samples captured under uncooperative scenarios, at a great distance, and under uncontrolled light conditions.

Other studies have analyzed biometric traits traditionally captured using touch-based sensors (e.g., fingerprint and hand characteristics) and have aimed at obtaining recognition systems that do not require the touch of a sensor or the use of placement guides.

This chapter is structured as follows. Section 3.1 presents a general description of the research trends of less-constrained biometrics. Section 3.2 presents a brief literature review of less-constrained biometric systems that traditionally perform the acquisition process using CCD cameras (the face, iris, gait, and ear, as well as soft biometric traits). Finally, Section 3.3 describes recent studies of touchless biometric systems based on hand characteristics, which are traditionally acquired using touch-based sensors.

3.1 Less-Constrained Biometric Systems

Most biometric systems require the acquisition process to be performed in a highly controlled manner. In these constrained applications, the user must perform deliberate actions to cooperate with the system. For example, many systems require that the user touch a surface, assume a specific pose, and remain motionless during the acquisition procedure.

In contrast, unconstrained biometric systems enable recognition in covert applications, without the subjects' knowledge, and in uncontrolled scenarios [90]. In these systems, biometric traits are captured by cameras that can adapt their positions and settings without requiring user cooperation. The algorithms employed can also compensate for non-idealities regarding noise, poses, and light conditions.

Researchers are studying different solutions to obtain unconstrained biometric technologies. These studies have analyzed techniques for reducing the constraints required by biometric acquisition procedures and have aimed at increasing the usability and user acceptance of recognition systems.

Studies regarding less-constrained biometric systems have been based on different biometric traits and their principal goals have included the following:

- ◼ Increasing the distance between the user and the sensor
- ◼ Reducing the required level of user cooperation
- ◼ Designing recognition methods compatible with uncontrolled light conditions
- ◼ Designing highly usable adaptive acquisition systems
- ◼ Designing preprocessing methods for reducing noise and enhancing data captured under less-constrained conditions
- ◼ Developing new feature extraction and matching algorithms specifically designed to obtain accurate results using data captured under less-constrained scenarios
- ◼ Designing methods that permit data compatible with existing biometric technologies to be obtained from samples captured under less-constrained applications

Some studies of less-constrained biometric techniques have been based on methods for computing and processing three-dimensional models, which can permit more information and less-distorted data to be obtained compared with two-dimensional acquisition techniques.

Many studies of less-constrained biometric systems have analyzed traits traditionally captured using CCD cameras (e.g., the face and iris). These traits are more adaptable for use in unconstrained applications than those that are usually captured by touch-based sensors (e.g., the fingerprint and hand shape). The goal of these studies has been to use cameras placed at long distances to perform unco-operative recognitions. However, in many situations, the accuracy of these systems can be insufficient. In fact, less-constrained face recognition systems suffer from problems due to different light conditions, poses, and aging. Systems based on iris images captured under less-constrained scenarios are subject to low visibility of the distinctive pattern, reflections, occlusions, and gaze deviation. Some studies have analyzed other biometric traits, such as gait and ear characteristics. Other studies have aimed at increasing recognition accuracy using soft biometric traits, which can also be adopted to perform continuous authentications and periodic reauthentications.

Biometric systems that use touch-based acquisition sensors are the most diffused in the literature. To reduce the constraints of the acquisition process, the primary goal is to design touchless recognition techniques that process images captured using CCD cameras. These traits do not permit the design of completely unconstrained systems because the users must display their body parts to the acquisition sensors. However, these traits are characterized by high distinctiveness and sufficient user acceptance.

Section 3.2 presents a brief literature review of less-constrained biometric systems based on traits captured using CCD cameras (the face, iris, gait, and ear, as well as soft biometric traits) is first presented. Finally, Section 3.3 presents touchless techniques designed for biometric characteristics of the hand.

3.2 Touchless Biometric Traits

Biometric traits traditionally captured using touchless sensors are suitable for use in unconstrained recognition systems. Many studies have examined the reduction of constraints in these systems. Two of the most researched technologies consist of less-constrained biometric systems based on the face and iris.

Biometric systems based on facial traits are characterized by high user acceptance and are based on CCD cameras that can be placed at long distances. Therefore, many studies of less-constrained biometric technologies are based on this trait. However, face recognition systems cannot provide sufficient accuracy for high-security applications. In contrast, the iris is usually considered the most accurate biometric trait and can be captured using touchless techniques based on CCD cameras.

Recent studies have also analyzed gait and ear traits. Other studies have proposed the use of soft biometric traits to increase the accuracy of less-constrained biometric applications.

3.2.1 Less-Constrained Face Recognition

Traditional face recognition systems produce satisfactory results under controlled conditions. For these systems, the samples consist of face images captured by a frontal camera under controlled light conditions. Moreover, the users must be cooperative to obtain good-quality acquisitions.

To overcome these limitations, researchers have recently begun to investigate face recognition under unconstrained conditions.

Reducing constraints can also increase the number of possible application contexts for biometric technologies based on the face trait. In fact, traditional biometric techniques are usually adopted for security applications (such as control to buildings, airports, border checkpoints, computer authentication, and biometric documents). Less-constrained biometric techniques based on the face trait can also be used in new scenarios.

- *Surveillance applications*: Biometric recognitions are performed using frame sequences captured by surveillance cameras without the subjects' knowledge and in uncontrolled scenarios. The used images present many non-idealities, such as low resolution, shadows, reflections, occlusions, and pose and expression variations. Recent studies have reported encouraging results [80,91] and have demonstrated that biometric recognitions can be performed at a distance of greater than 15 m [92]. Biometric systems for surveillance applications can also be integrated with methods for understanding human behaviors [93].
- *Mobile devices*: Face recognition techniques can be integrated in mobile devices such as mobile phones [94,95]. This type of applications requires the use of fast algorithms robust to noise.
- *Multimedia environments with adaptive human–computer interfaces*: The use of less-constrained acquisition techniques can permit the design of new multimedia environments (e.g., behavior monitoring at childcare, assisted living centers, or video games) [96,97].
- *Video indexing*: Biometric techniques are used to label faces in frame sequences captured under different environments and light conditions [98].

Figure 3.1 presents examples of face recognition techniques applied in surveillance applications (Figure 3.1a), video indexing (Figure 3.1b), video games (Figure 3.1c), and mobile phones (Figure 3.1d).

As reported in [101], important aspects that must be considered in the design of less-constrained face recognition systems are variations in light conditions, poses, and aging.

- *Illumination*: The different techniques to compensate for illumination variations can be grouped into three principal classes: subspace methods, reflectance-model methods, and methods based on three-dimensional models. Subspaces methods are commonly used in the literature and can capture the generic face space, thus recognizing new samples not present in the training sets. Studies have examined the improved robustness of traditional subspace methods to variations in light conditions. For example, Turk and [102] reported that the Eigenface method is more robust to different illuminations if the first three eigenvectors are not considered. The second class

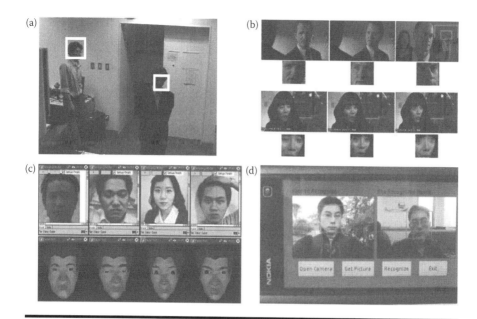

Figure 3.1 Examples of less-constrained face recognition applications: (a) Surveillance, (b) video indexing, (c) video games, and (d) mobile phones. ((a) Y. Ishii et al., Face and head detection for a realtime surveillance system, in *Proceedings of the 17th International Conference on Pattern Recognition*, vol. 3, pp. 298–301. © 2004, IEEE. (b) J. Y. Choi, W. De Neve, and Y. M. Ro, Towards an automatic face indexing system for actor-based video services in an IPTV environment, *IEEE Transactions on Consumer Electronics*, vol. 56, no. 1, pp. 147–155. © 2010, IEEE. (c) S. Chin and K.-Y. Kim, Expressive 3D face for mobile devices, *IEEE Transactions on Consumer Electronics*, vol. 54, no. 3, pp. 1294–1302. © 2008, IEEE. (d) B. Chen, J. Shen, and H. Sun, A fast face recognition system on mobile phone, in *Proceedings of the International Conference on Systems and Informatics*, pp. 1783–1786. © 2010, IEEE.)

of methods employs a Lambertian reflectance model with a varying albedo field. An example of reflectance-model methods is reported in [103]. Methods based on three-dimensional models are usually more robust to illumination variations, but require more complex data and algorithms. Examples of these methods are the Eigenhead [104] and the morphable model approach [105].

■ *Poses*: Traditional biometric systems may yield poor results for face images captured using uncontrolled poses. Thus, pose differences must be compensated using dedicated methods. Pose compensation can be considered a correspondence problem because this compensation aims to estimate the face position in three-dimensional space [101]. Different methods of compensating for pose variations in face recognition systems have been described. Many methods aim to recover the three-dimensional shape of the face from two-dimensional images [101,106]. The relevant is reviewed in [107].

■ *Aging*: The aging effect is particularly important in unconstrained recognition systems that require long-time enrollments, such as surveillance, investigative, and forensic applications. Different techniques for compensating for aging have been described. Many methods simulate aging effects [108], and other techniques use matching strategies specifically designed to be robust to age variations [109].

Some less-constrained face recognition systems use frame sequences and multiple views to increase accuracy by evaluating the information related to temporal continuity [110] and to three-dimensional shape [111].

3.2.2 Less-Constrained Iris Recognition

Iris recognition is considered one of the most accurate biometric techniques. However, iris recognition systems are usually based on complex and expensive acquisition techniques because the iris region of the eye is a relatively small area that is wet and constantly in motion due to involuntary eye movements. Moreover, the iris can be occluded by eyelids, eyelashes, glasses, and reflections.

Most iris acquisition systems require that the user remain motionless watching a camera placed at a short distance from the eyes (usually less than 30 cm). Therefore, the acquisition process is time-consuming and can require several trials to obtain sufficient quality biometric samples. Moreover, iris acquisition systems are usually based on near-infrared illumination techniques, which can be perceived as dangerous to health. These aspects drastically influence the usability and user acceptance of iris recognition systems.

To overcome these problems, researchers are studying less-constrained iris recognition systems. The primary goals are to increase the distance between the iris and the sensor, to use images captured on the move, and to use iris samples acquired under uncontrolled light conditions. An example of an iris recognition system based on images captured at a distance on the move is presented in Figure 3.2.

Sun et al. [112] classified iris recognition systems into seven categories based on three characteristics: the techniques used to search the iris region, the distance between the eye and the sensor, and the required level of cooperation during the biometric acquisition.

1. *Close-range iris recognition*: This category of biometric systems is the most diffused in the literature. The eye must be placed at a small distance from the camera, and the user must remain motionless watching the camera during the acquisition process.

2. *Active iris recognition*: The iris images are captured at a small distance from the camera; however, the user position is less constrained because active cameras reduce the required level of cooperation. In fact, the recognition systems can automatically detect the iris position and align the camera with the iris. These

Figure 3.2 Example of an iris recognition system based on images captured at a distance on the move. (J. Matey et al., Iris on the move: acquisition of images for iris recognition in less constrained environments, *Proceedings of the IEEE*, vol. 94, no. 11, pp. 1936–1947. © 2006, IEEE.)

systems usually capture a face image using a wide-angle camera, estimate the iris coordinates, and then move the iris camera toward the iris region [113].

3. *Iris recognition at a distance*: These systems use passive cameras but can capture iris images at a long distance. The user must be cooperative and remain motionless during the biometric acquisitions. Techniques used to perform iris recognition at a distance were investigated in [114].

4. *Active iris recognition at a distance*: This category of iris recognition systems permits good-quality samples to be obtained with a reduced level of cooperation compared to iris recognition at a distance. However, the users must remain motionless in a fixed position. An example of these systems is proposed in [115].

5. *Passive iris recognition on the move*: Iris images are captured on the move in these systems. However, the use of passive cameras requires predefined walking paths. An example of these systems is the biometric portal described in [81], which is based on multiple passive cameras.

6. *Active iris recognition on the move*: Active cameras can search the irises of people walking in a defined direction, reducing constraints and costs compared with passive iris recognition on the move. No examples of these systems have been described.

7. *Iris recognition for surveillance*: Active iris camera networks can capture samples from multiple individuals. The acquisition process is completely unconstrained. In the future, this type of biometric system will likely be used in a wide range of application contexts that require accurate identification techniques (e.g., investigative and governmental applications).

Important research topics regarding the design of less-constrained iris recognition systems include segmentation algorithms, recognition techniques, and gaze assessment methods.

■ *Segmentation*: The task that locates and separates the iris pattern in the input face/eye image is called segmentation [7,116]. This task is particularly critical for less-constrained recognition systems because the captured images can present more reflections and occlusions compared with traditional iris samples. Moreover, the correct detection of the iris boundaries and the removal of occlusions are directly related to the accuracy of the iris recognition system. Different methods specifically designed for estimating the iris boundaries in iris images captured using less-constrained techniques have been described. These methods can be based on computational intelligence techniques [117,118], active contours [119], or incremental approaches [32]. Then, reflections and occlusions are searched using statistical approaches [116] or more complex techniques [120–122].

■ *Recognition techniques*: Images captured under less-constrained conditions present lower visibility of the ridge pattern compared with images captured by traditional iris acquisition systems. Thus, dedicated techniques must be used for enhancement, feature extraction, and matching. For example, the method proposed in [123] is specifically designed for frame sequences captured by a biometric portal and uses super-resolution techniques to enhance the visibility of the ridge pattern. Results obtained using different matching methods for images captured at a distance, on the move, and under natural light conditions are reported in [79]. Algorithms specifically designed for mobile applications have also been described [124].

■ *Gaze assessment*: Iris images are subject to gaze deviation with respect to the camera. Off-axis images can drastically reduce recognition accuracy because these images can affect the performance of the segmentation and feature extraction algorithms. The effect of this non-ideality can be reduced using methods for gaze assessment [125] or acquisition setups based on multiple cameras [126].

3.2.3 Soft Biometrics

Although featuring a lack of distinctiveness, soft biometric recognition techniques can employ samples captured in an unobtrusive and unconstrained manner under uncooperative conditions [20,23,35] or with surveillance cameras placed at long distances [23]. These recognition systems can be employed when adopting systems based on hard biometric traits (e.g., surveillance applications) is difficult, when the pool of users is sufficiently small, or when high accuracy is not required.

Soft biometric traits can be used in different application contexts. The approach presented by [127] uses a set of comparative labels to describe soft biometric characteristics and to perform biometric recognitions. The technique proposed by [24] aims to reduce the number of entries to be analyzed in biometric queries performed on large surveillance databases by evaluating categorical information such as gender and race. A continuous authentication technique based on face and clothing is described in [128]. Color and height characteristics are used in [21] to detect the individuals in a multicamera network. A recognition method based on a three-part (head, torso, and legs) height and color soft biometric model is presented in [23]. Algorithms for computing different soft biometric features (gait, height, size, and gender) are proposed in [129]. The method described in is designed for working under unconstrained conditions and can estimate the weight of walking individuals from surveillance frame sequences. This technique can be particularly useful in forensic analyses because weight is one of the few characteristics that can be inferred from scene evaluation.

3.2.4 Other Biometric Traits

Other biometric traits suitable for unconstrained recognition applications include gait and ear shape.

Gait characteristics are estimated from frame sequences captured by CCD cameras [18]. Gait recognition systems can obtain satisfactory accuracy and can work at great distances [130] and with poor-quality frame sequences [131]. However, gait is a behavioral biometric trait; therefore, gait recognition systems feature lower accuracy compared with biometric systems based on physiological characteristics.

Biometric systems based on ear shape are recent technologies [13,14]. Studies have examined less-constrained recognition systems based on the ear trait [132,133]. However, less-constrained ear recognition systems are less accurate than more mature biometric recognition technologies.

3.3 Touch-Based Biometric Traits

Biometric systems based on hand characteristics traditionally require the use of touch-based acquisition sensors. These systems can be based on the evaluation of hand shape [10], palmprint [11], palm-vein [12], or multiple traits [134].

Most hand biometric systems use acquisition sensors composed of a flat surface and pegs that guide the placement of the user's hand. Recent studies have analyzed techniques for reducing the constraints imposed by acquisition technologies. Based on the acquisition technique, reported hand recognition approaches can be classified into the following three categories [135]:

- Constrained and touch-based systems that use pegs or pins to constrain the position and posture of the hand.
- Unconstrained and touch-based systems that do not use pegs and permit a less-constrained placement of the hand on the acquisition sensor. These systems often require the user to place the hand on a flat surface [136] or on a digital scanner [137]. These systems are characterized by increased

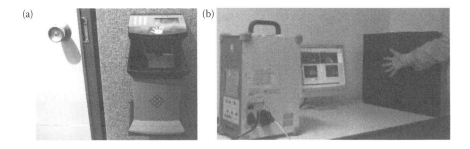

Figure 3.3 Examples of palmprint acquisition sensors: (a) a constrained and touch-based system and (b) an unconstrained and touchless system. ((a) M. Wong et al., Real-time palmprint acquisition system design, *IEEE Proceedings on Vision, Image and Signal Processing,* **vol. 152, no. 5, pp. 527–534. © 2005, IEEE. (b) V. Kanhangad, A. Kumar, and D. Zhang, A unified framework for contactless hand verification,** *IEEE Transactions on Information Forensics and Security,* **vol. 6, no. 3, pp. 1014–1027. © 2011, IEEE.)**

user acceptance compared with constrained and touch-based systems. However, these systems require more complex strategies to perform sample alignment [10].

■ Unconstrained and touchless systems use CCD cameras to capture hand images. These systems do not require the contact of the hand on acquisition surfaces. Many reported systems are based on single images [138–140]; however, biometric systems that use three-dimensional models also exist [135,138,141]. Most of the unconstrained and touchless systems are designed to work under controlled light and background conditions [142]. Other systems do not impose these constraints and use more complex segmentation strategies [143]. Systems integrated in mobile devices have also been examined [144].

Figure 3.3 presents examples of a constrained and touch-based system [145] and of an unconstrained and touchless system [146].

3.4 Summary

The acquisition techniques traditionally adopted by biometric systems require users to perform deliberate actions to cooperate with the system and impose constraints. For example, many systems require that the user touch a surface, assume a defined pose, and remain motionless during the acquisition procedure.

An important goal of the international research community is the design of unconstrained biometric applications in which the recognition process is performed covertly and without the user's knowledge.

Many studies have focused on reducing constraints in biometric systems. The primary goals of these studies include increasing the distance between the users and the system, designing recognition techniques compatible with uncontrolled light conditions, increasing the usability and user acceptance of the systems, designing feature extraction and matching techniques robust to noise, and implementing techniques that guarantee the compatibility of the templates computed by less-constrained systems with existing biometric databases.

Biometric characteristics traditionally acquired using touchless techniques (e.g., face, iris, gait, and ear, as well as soft biometric traits) are more suitable for use in unconstrained applications than traits traditionally captured using touch-based sensors (e.g., fingerprint and hand characteristics). The primary goal of studies of less-constrained biometric systems based on the first class of traits is the use of CCD cameras placed at long distances to perform uncooperative recognitions. Less-constrained face recognition systems can be applied in new application contexts, such as surveillance applications, mobile devices, multimedia environments with adaptive human–computer interfaces, and video indexing. However, the design of less-constrained face recognition techniques must consider important non-idealities, such as uncontrolled illumination conditions, pose variations, and aging. Biometric systems based on the iris trait are typically more accurate than face recognition systems but are usually based on complex and expensive acquisition techniques. Recent studies of iris recognition systems have aimed at designing techniques compatible with images captured at long distances, on the move, and under uncontrolled light conditions. Important research topics regarding the design of less-constrained iris recognition systems include segmentation algorithms, recognition techniques, and gaze assessment methods. Other characteristics that can be acquired in unconstrained scenarios include soft biometric traits, which can also be used to perform continuous authentications and periodic reauthentications. Studies have also examined less-constrained biometric systems based on gait and ear traits.

Biometric systems that traditionally use touch-based acquisition sensors are the most diffused in the literature. Therefore, researchers are studying less-constrained recognition systems, with particular attention to methods based on hand characteristics. Depending on the acquisition technique used, hand recognition approaches described in the literature can be divided into the following three categories: constrained and touch-based systems, unconstrained and touch-based systems, and unconstrained and touchless systems. The third category of methods performs acquisitions at a distance and can use two-dimensional or three-dimensional data.

Chapter 4

Fingerprint Biometrics

The fingerprint is one of the most commonly used traits in biometric applications due to its high durability and distinctiveness.

Fingerprint sample analysis can be performed at the global (Level 1), thin (Level 2), and ultra-thin (Level 3) levels. Many studies have examined techniques based on different analysis levels and designed to perform every step of the biometric recognition process. Typically, these steps include acquisition, quality evaluation, enhancement, feature extraction, and matching. Other methods have been proposed for classifying and indexing fingerprint samples in identification applications and for computing synthetic fingerprint images that can be used to design and test biometric algorithms.

Most fingerprint recognition systems use acquisition sensors that require contact of the finger with a platen. However, these systems suffer from problems due to the contact of the finger with a surface (e.g., distortions in the captured images and latent fingerprints on the acquisition surface). To overcome these problems and increase user acceptability of the biometric recognition process, touchless recognition systems are being studied. These systems are based on CCD cameras and can be classified into systems based on two-dimensional and three-dimensional samples. Most of these systems aim to guarantee compatibility with the existing AFIS.

4.1 Fingerprint Recognition

Among possible biometrics, fingerprints are the most well known and widely used in recognition applications. Their use in police investigation can be traced back to the late nineteenth century, and the first automatic fingerprint recognition system was introduced in the 1970s [147].

Fingerprint recognition systems are based on the analysis of the impression left by the friction ridges of the human finger. The evaluation of the ridge details can be performed with three different levels of accuracy:

- Level 1: The overall global ridge flow pattern is considered.
- Level 2: The analysis is based on distinctive points of the ridges, which are called minutiae points.
- Level 3: Ultrathin details, such as pores and incipient ridges, are studied.

These analysis levels can be used in different modules of biometric recognition systems. For example, Level 1 characteristics can be used to perform preliminary tasks, such as the quality evaluation of biometric samples and the enhancement of ridge pattern visibility. Then, more detailed analyses can be used to perform identity comparisons.

Fingerprint samples are typically acquired using touch-based techniques. These techniques can be based on different technologies and can be used in live-recognition systems or in forensic analyses.

In general, fingerprint recognition systems can be divided into the following modules: acquisition, quality evaluation, enhancement, feature extraction, and matching. Moreover, classification and indexing techniques in identification systems can be used to reduce the number of identity comparisons performed for each biometric query. Many different methods have been specifically designed for each step of the recognition process based on fingerprint images captured using touch-based techniques. Algorithms for computing synthetic fingerprint images have also been designed; these algorithms can be used to reduce the efforts necessary to capture biometric samples during the design of recognition systems.

Touch-based fingerprint recognition systems can obtain remarkable accuracy but suffer from important problems, such as filth on the acquisition surface, distortions in captured images due to elastic skin deformations, and the potential of obtaining latent fingerprints from the surface of the sensors. Recent studies have aimed at overcoming these problems by using touchless acquisition techniques. Touchless fingerprint recognition systems can be divided into two classes: systems based on two-dimensional samples and systems based on three-dimensional samples. The first class of biometric systems usually captures single fingerprint images with CCD cameras, whereas the second requires hardware setups that are more complex but can provide greater accuracy.

In this chapter, the characteristics of the fingerprint biometric trait (Section 4.2) and its application contexts are first discussed (Section 4.3). Then, the techniques designed for the analysis of fingerprint images at different levels are discussed in Section 4.4. Next, touch-based fingerprint recognition systems are reviewed in Section 4.5, including the image capture, quality evaluation techniques, image enhancement algorithms, feature extraction and matching methods, classification and indexing approaches, and algorithms for computing synthetic fingerprint images. Section 4.6.1 describes recognition techniques based on touchless two-dimensional fingerprint samples. Finally, Section 4.6.2 analyzes three-dimensional reconstruction techniques and recognition methods used by biometric systems based on touchless three-dimensional samples.

4.2 Characteristics of the Fingerprint

The fingerprint is a highly durable biometric trait [148]. The fingertip ridge structure is fully formed at approximately the seventh month of fetal development, and this pattern configuration does not change throughout the lifetime of the individual unless serious accidents or diseases occur [149]. Cuts and bruises usually modify the fingerprint pattern only temporarily. Thus, the fingerprint is an attractive biometric trait, particularly for application contexts that require long-term enrollment.

Another important property of the fingerprint is its uniqueness. In general, fingerprints are part of an individual's phenotype and are different for different individuals. In addition, the fingerprints of the same person are different, even between identical twins [150,151].

However, the uniqueness of fingerprints is not an established fact but an empirical observation [5]. Studies have examined the amount of distinctive information encoded in the fingerprint pattern. [152] reported that the distinctiveness of a fingerprint is proportional to the number of minutiae in the considered image, which is related to the size of the available fingerprint portion. Other studies have compared the distinctiveness of different features extracted from the fingerprint pattern. For example, [153] compared Level 1, Level 2, and Level 3 features. However, estimating fingerprint distinctiveness is an open problem and is particularly important in legislative and forensic applications [154].

4.3 Applications

The fingerprint trait is the most commonly used and known biometric characteristic [28].

The applications of fingerprint recognition technologies are heterogeneous and range from the public to the private sector. Commercially available fingerprint recognition systems vary widely in sizes of sensors, costs, and accuracy [155]. For example, fingerprint recognition systems have been integrated in electronic devices (e.g., PDA and mobile phones [156]), on-card systems [157], systems based on a single personal computer, and large distributed systems, such as AFIS [29].

The primary application contexts of fingerprint recognition systems are in the forensics, government, and commercial sectors [5]. In the forensics sector, the fingerprint trait is used for identification, missing person searches, and general investigative activities. In the government sector, important applications of the fingerprint trait include border control and biometric documents (e.g., passports

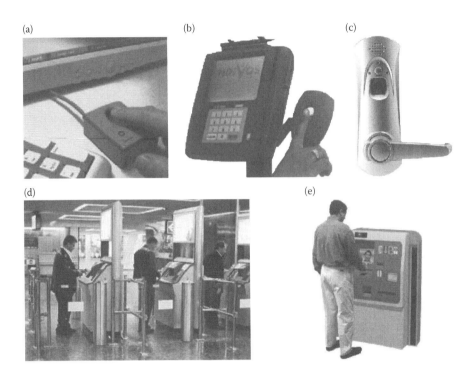

Figure 4.1 Examples of applications based on the fingerprint trait. (a) personal computer, (b) point of sale (POS) terminal, (c) access to protected areas, (d) border control, and (e) automatic teller machine (ATM). (A. Jain, A. Ross, and S. Prabhakar, An introduction to biometric recognition, *IEEE Transaction on Circuits and Systems for Video Technology*, vol. 14, no. 1, pp. 4–20. © 2004, IEEE.)

and IDs). Examples of applications in the commercial sector include authentication systems integrated into ATMs, terminal login, access control for on-line services (e.g., e-commerce and e-banking applications), protection of sensitive data (e.g., in personal computers, PDA, mobile phones, and storage devices), and access control to restricted areas.

Figure 4.1 presents some examples of applications based on the fingerprint trait.

4.4 Analysis of Fingerprint Samples

This section presents reported techniques for analyzing touch-based and touchless fingerprint images.

4.4.1 Level 1 Analysis

Techniques for the Level 1 analysis of fingerprint images evaluate the overall ridge flow. Examples of characteristics analyzed at Level 1 are the ridge orientation, local ridge frequency, singular regions, and ridge count.

Figure 4.2   Example of a ridge orientation map.

The local ridge orientation is estimated as the angle of the ridges with respect to the horizontal axis. Usually, the orientation angle is considered an unoriented direction lying in $0°, \ldots, 180°$. The ridge orientation map can be computed by estimating the local ride orientation in areas centered in each pixel of the fingerprint images or by using techniques based on global orientation models. Figure 4.2 presents an example of a ridge orientation map. Most reported methods for computing the ridge orientation are based on local analysis techniques [158]. These techniques are usually adopted by fingerprint enhancement methods. One of the best-known local analysis techniques is based on evaluating the gradient orientation in squared regions. For each local region, the ridge orientation is represented by the mean of the angular values of the pixels. This technique was first introduced in [159]. One of the primary problems of this technique is the difficulty in obtaining robust means of the angular values. Some methods based on gradient analysis use different algorithms for computing the mean angle [160,161]. Other local techniques are based on slit-based approaches [162] or on the frequency analysis of local regions of the fingerprint image [163,164]. Orientation regularization techniques in the literature have been designed to reduce artifacts due to the presence of noise or poor-quality image regions [165,166]. Global approaches are usually adopted during the enhancement of latent fingerprints or for the computation of synthetic fingerprint images. The method described in [167] is based on a mathematical model that estimates the local ridge orientation by considering the coordinates of core and delta points.

The local ridge frequency is the number of ridges per unit length along a segment orthogonal to the ridge orientation. Many reported methods assume that the ridge frequency of a fingerprint image can be represented using a single value. Other methods compute a ridge frequency map, which represents the local ridge frequency computed in each pixel of the fingerprint image. One of the most commonly used algorithms in the literature for estimating the ridge frequency map is based on computing local oriented x-signatures [168]. This algorithm divides the image in local windows centered in every pixel with a fixed size, and the orientation is computed according to the ridge orientation map. Then, the x-signature of every window is obtained as a vector containing a cumulative function of the intensity values computed for each column x. The frequency is finally estimated as the inverse of the average distance between consecutive peaks of the image obtained by computing the x-signature. Other algorithms based on the concept of an x-signature have been proposed to reduce the sensitivity to the noise of the method [169,170]. Methods for estimating the ridge frequency have been designed based on different techniques, such as the short-time Fourier transform (STFT) [164] and local curve region analyses [171].

In another important Level 1 analysis, singular regions are estimated. A singular region represents an area of the finger with a distinctive ridge shape. Commonly, three different types of singular regions are considered: the loop, delta, and whorl. The distinctive shapes of these regions are ∩, Δ, and O, respectively (Figure 4.3). The majority of reported methods for estimating the singular regions use the Poincaré technique [172]. Figure 4.4 presents the schema of this method. The Poincaré index is computed for each point of a ridge orientation map. This index represents the total rotation along the considered pixel. The value of the Poincaré index indicates the presence of a singular region. The value 180 corresponds to a loop, −180 represents a delta, and 360 indicates a whorl. Other reported methods for detecting singular regions are based on analyzing the ridge orientation map but use different analysis techniques. For example, the method proposed in [173] estimates the orientation of the ridges using an algorithm based on the

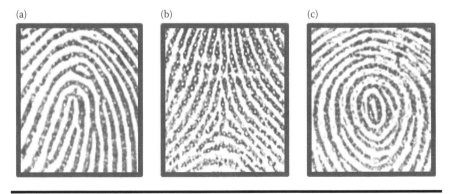

Figure 4.3 **Examples of singular regions: (a) loop, (b) delta, and (c) whorl.**

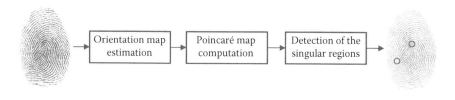

Figure 4.4 Schema of the Poincaré technique for detecting singular regions in fingerprint images.

Zero-pole Model, and the method described in [174] is based on analyzing the topological structure of the fingerprint image.

Many biometric fingerprint systems compare fingerprint images using a specific reference point of the image called the core point. This point is usually estimated using Level 1 analysis techniques. A simple technique consists of selecting the northern loop [5]. However, many reported methods use more complex techniques that are based on searching singular regions [175]. Other methods use different strategies based on analyzing the ridge orientation. The study [176] considered the core as the point in the middle of two center-of-gravity points obtained by computing the local axial symmetry of the image. [177] quantized the directional map of the ridges in four different binary images related to the four cardinal points. Then, the reference point is obtained by applying a set of rules and morphological operators to these images. The method described in [178] is specifically designed for fingerprint images that do not present singular regions. The local orientation of the ridges is computed, and then the radial symmetry line is estimated. The core consists of the point with the maximum value in the radial symmetry line.

Examples of other Level 1 characteristics include the ridge count [179] and other global mapping information obtained by applying Gabor filters to the input fingerprint images [180]. The ridge count is an abstract measurement of the distances between any two points in a fingerprint image. This characteristic is frequently used in forensic analyses performed by human operators. In fact, a typical measurement performed by forensic experts is the number of ridges between two singular regions. In contrast, Gabor filters are used to extract information related to the ridge frequency and orientation in local regions of the fingerprint images.

4.4.2 Level 2 Analysis

Level 2 analysis evaluates specific ridge discontinuities called minutiae. Different classes of minutiae can be distinguished (Figure 4.5); however, most published automatic biometric systems only consider terminations and bifurcations.

Many studies have examined techniques for extracting minutiae points [181]. These methods can be based on the computation of the binary images of the ridge pattern or the direct extraction of minutiae from gray-scale images. Methods based

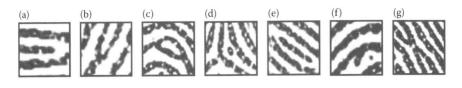

Figure 4.5 **Examples of common minutiae types: (a) termination, (b) bifurcation, (c) lake, (d) point or island, (e) independent ridge, (f) spur, and (g) crossover.**

on computing binary images can require processing of thinned images representing the skeleton of the ridges or direct searching of the minutiae in the binary images. Figure 4.6 presents a classification of published methods for minutiae extraction [181].

The most diffused techniques extract minutiae points from thinned images. These techniques can be divided into the following four primary steps [5]:

1. Adaptive binarization to separate the ridges from the background of the fingerprint image.
2. A thinning operation to reduce the thickness of the ridges to a single pixel.
3. Estimation of the coordinates of the minutiae by observing the specific local pattern of each single pixel of the ridges, typically in its 8-neighborhood.
4. A postprocessing method to reduce the number of false minutiae detected in the previous step.

Figure 4.7 presents the schema of the described method.

Many techniques have been designed for fingerprint image binarization. The simplest technique consists of the use of a global threshold [182]. However, the obtained results can present artifacts and noisy regions due to differences in finger pressure on the sensor. Another simple technique used by the "FBI minutiae reader" is based on a local thresholding approach and on an algorithm that compares the pixel alignment along eight discrete directions [183]. Other techniques are based on different approaches, for example, fuzzy logic [184], directional filters tuned

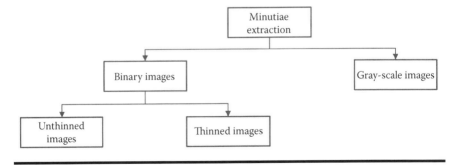

Figure 4.6 **Classification of the minutiae extraction techniques.**

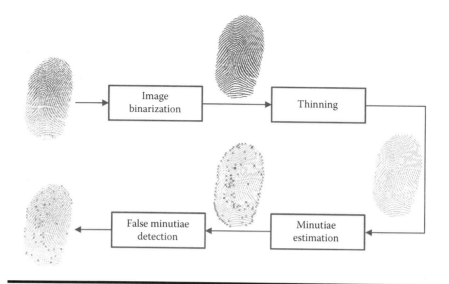

Figure 4.7 Schema of the most diffused minutiae extraction methods.

according to the ridge frequency [185], and iterative algorithms that follows the shape of the ridges [186]. The second step is computing the ridge skeleton.

One of the most commonly used techniques for thinning binary images representing the ridge pattern is based on morphological operators [182]. To limit problems due to artifacts introduced by morphological operators, some techniques have been specifically designed for analyzing the ridge skeleton and for removing false ridges [187]. Other techniques are based on different approaches. For example, the technique described in [188] uses pulse-coupled neural networks, and the algorithm presented in [189] is based on analyzing the ridge orientation.

Then, the coordinates of the minutiae points are estimated. A well-known algorithm for detecting minutiae points in thinned fingerprint images is based on computing the crossing number [190]. This value is computed by evaluating every pixel n_k in the eight neighboring pixels of every point p as follows:

$$\mathrm{CN}(p) = \frac{1}{2} \sum_{k=1}^{8} |n_k - n_{((k+1)\,\mathrm{mod}\,8)}|, \qquad (4.1)$$

where $n_k \in \{0, 1\}$. Terminations are characterized by a crossing number $\mathrm{CN}(p) = 1$, and bifurcations have a value $\mathrm{CN}(p) = 3$. Other methods search the coordinates of the minutiae points in thinned images using different algorithms, such as morphologic operators [191].

The final step is the refinement of the obtained results to remove false minutiae. Some techniques are based on analyzing thinned fingerprint images or evaluating the shape of the ridges in gray-scale images. The first class of methods searches local patterns that describe false minutiae introduced by the thinning step [186,192]. The methods based on evaluating gray-scale images extract a set of features for each minutiae point and then classify every minutia as valid or false. A method for this class based on neural classifiers is proposed in [193].

To limit the introduction of artifacts by the thinning step, many algorithms search the minutiae points in the binary image representing the ridge pattern without computing the ridge skeleton. For example, the software NIST MINDTCT [186] is based on a local search of all binary patterns that define the presence of terminations and bifurcations. The method uses a set of 10 binary masks with 3×2 pixels and scans the binary fingerprint image in the vertical and horizontal directions. Reference [194] presented another method based on local analysis of the ridge pattern and on examining the intensity along squared paths in the image. Other methods use run-based algorithms [195], which search the minutiae by analyzing the graphs obtained by computing the horizontal and vertical run-length encoding of the binary fingerprint image. Another technique based on analyzing binary images representing the ridge pattern searches the ridges by analyzing the chaincode obtained from the fingerprint image [196]. The chaincode is a reversible encoding technique for binary images. For each object in a binary image, the information related to its boundary is stored in a vector. Starting from an initial position, the boundary is followed by a searching algorithm, and the angles of the sequent pixels are stored. The minutiae are estimated by searching direction changes in the chaincode vectors.

Other approaches directly search the coordinates of the minutiae points in gray-scale images. A well-known method based on a ridge-following technique is described in [197]. An iterative algorithm follows every ridge according to the angle described by the ridge orientation map. The algorithm stops when a termination or bifurcation is detected. Variants of this method also consider the two valleys near every ridge [198] or are designed for low-cost hardware devices [199]. Other methods that search the minutiae in gray-scale images use different strategies. Neural network classifiers are used in [200], a local searching algorithm based on fuzzy logic is described in [201], and an approach based on computing the linear symmetry is proposed in [202].

Most reported techniques also estimate the orientations of the minutiae based on the minutiae coordinates in the ridge orientation map. Many techniques also estimate the correctness probability of every minutia as the local quality value of the fingerprint image [186].

4.4.3 Level 3 Analysis

Level 3 analysis requires high-resolution acquisition devices (with at least 800 ppi [203]) and is not commonly applied in commercial systems. Details that

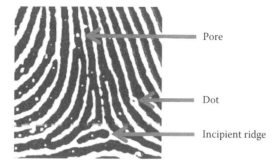

Figure 4.8 Example of fingerprint details considered during Level 3 analysis: pores, dots, and incipient ridges.

can be considered at this level of analysis include pores, dots, and incipient ridges. Figure 4.8 presents an example of Level 3 characteristics. Usually, the features extracted by automatic algorithms at this level of analysis consist of the spatial coordinates of the pores. Recent studies have demonstrated that biometric recognition techniques based on Level 3 features can obtain greater accuracy than recognition techniques based on Level 1 and Level 2 features [204]. Level 3 features can also be used for vitality detection during touch-based fingerprint acquisitions [205].

4.5 Touch-Based Fingerprint Recognition

Biometric systems based on the fingerprint trait estimate the identity of an individual by extracting and comparing information related to the characteristics of the ridge pattern.

The schema of the biometric authentication process based on the fingerprint trait is presented in Figure 4.9. The first step is acquiring the biometric sample. The sample consists of an image that can be captured using different types of sensors. In most fingerprint recognition systems, a quality evaluation of the captured image is then performed to discard samples of insufficient quality. Another typical step applies specifically designed techniques to enhance the visibility of the ridges. The next step is feature extraction, which computes a biometric template from the captured fingerprint image. Some reported recognition methods are based on different characteristics of the ridge pattern. Then, the obtained template is compared with the stored data during the matching step. The algorithms used are strictly dependent on the considered features. Finally, a decision is obtained by applying a threshold to the obtained match score.

Biometric systems based on the fingerprint trait estimate the identity of an individual by extracting and comparing information related to the characteristics of the ridge pattern.

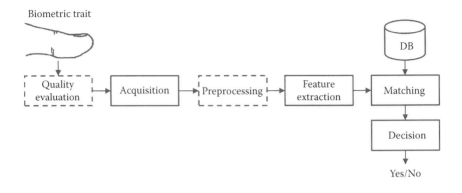

Figure 4.9 **Schema of the biometric authentication process based on the fingerprint trait.**

Moreover, many fingerprint recognition systems designed for identification perform a classification or indexing step to reduce the number of biometric queries.

The design of fingerprint recognition systems requires the collection of large datasets of samples. This task is expensive and time-consuming and can require implementing new hardware setups. Thus, methods have been specifically designed to compute realistic synthetic images, which can be used to test the envisioned methods and to reduce the efforts necessary to collect a sufficient number of biometric samples.

This section reviews the most important methods for the different steps of the biometric recognition process based on the fingerprint trait.

4.5.1 Acquisition and Fingerprint Images

Based on the acquisition technique used, fingerprint images can be divided into three classes: latent fingerprints, inked fingerprints, and live-scan fingerprints. Figure 4.10 presents examples of images obtained using the different fingerprint acquisition methods.

■ Latent fingerprints are extremely important in forensics. This type of fingerprint is produced by transferring the film of moisture and grease that is present on the finger surface to an object that is touched. Latent fingerprints are usually not visible to the naked eye, and forensic investigators use appropriate reagents to enhance the visibility of the ridge pattern [206].
■ Inked fingerprints are typically obtained as follows. First, black/blue ink is spread on the user's finger, and the finger is rolled on a paper card. Second, the card is converted into a digital form using a high-definition paper-scanner or a high-quality CCD camera [207].

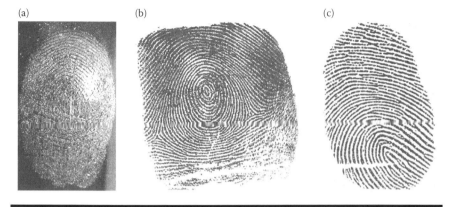

Figure 4.10 Examples of fingerprint images: (a) latent, (b) rolled and inked, and (c) live-scan.

■ Live-scan fingerprints are obtained by impressing a finger on the acquisition surface of a device. Sensors on the device capture images of one or more fingerprints. Two types of live-scan sensors can be distinguished: area scan sensors and swipe sensors [208]. Examples of area scan sensors and swipe sensors are presented in Figure 4.11. The first category of sensors permits the fingerprint pattern to be captured in a single time instant. Swipe sensors require the user to slide a finger vertically over the surface. These sensors are smaller and cheaper than area scan sensors, but require previous user training and

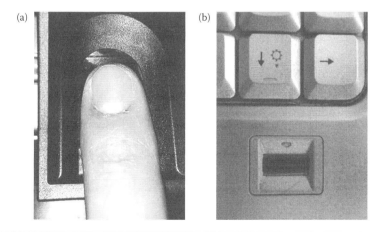

Figure 4.11 Examples of fingerprint acquisition sensors: (a) area scan sensor and (b) swipe sensor.

often fail to capture fingerprint samples. The sensors can be based on different technologies, including optical, capacitive, thermal, piezoelectric, radio frequency, ultrasonic, microelectromechanical (MEMS), and multispectral imaging [208].

Only good-quality images must be stored to achieve subsequent accurate biometric recognitions by automatic systems. For example, the Federal Bureau of Investigation (FBI) of the United States has defined a set of primary parameters characterizing the acquisition of a digital fingerprint image [209] that include the minimum resolution, minimum size of the captured area, minimum number of pixels, maximum geometry distortion of the acquisition device, minimum number of gray levels, maximum gray-level uniformity, minimum spatial frequency response of the device, and minimum signal-to-noise ratio.

In many law enforcement and government applications that involve AFIS, the size of the fingerprint image database is extremely large. For example, the FBI fingerprint card archive contains over 200 million identities [210]. In such cases, compressed formats are adopted to store the biometric samples. Most reported image compression algorithms produce unsatisfactory results for compressing fingerprint images. One of the most used compression algorithms had been proposed by the FBI and is based on wavelet scalar quantization (WSQ) [211]. This algorithm computes the scalar quantization of a 64-subband discrete wavelet transform decomposition of the image, followed by Huffman coding. Another diffused compression technique is JPEG 2000 [212], which is also adopted for different types of images. The most used compression algorithms are compared in [213].

4.5.2 Quality Estimation of Fingerprint Samples

The level of quality of captured fingerprint images can vary widely. Low levels of fingerprint quality can compromise recognition accuracy [62]. Quality estimation methods are usually considered to control this factor [214]. Quality estimation is also useful for selecting unrecoverable image regions and for properly weighing the extracted features according to the local quality level of the input fingerprint image.

Examples of fingerprint images affected by different non-idealities are presented in Figure 4.12.

Many quality evaluation techniques designed for touch-based fingerprint images are based on an evaluation of local features. The method described in [216] evaluates the probability density function (PDF) of local regions. The technique presented in [217] computes the features used by applying a set of Gabor filters with different orientations to the local areas of the image. Other techniques are based on global characteristics. For example, the method described in [218] is based on an evaluation of the energy distribution rate in wavelet-compressed fingerprint images. The technique proposed in [219] combines local and global features related to the frequency domain (a ring structure of DFT magnitude and directional Gabor features) and to the spatial domain (black pixel ratio of the central area). In addition, the

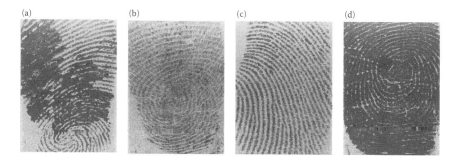

Figure 4.12 **Examples of fingerprint images affected by different non-idealities. (a) a fingerprint covered with liquid; (b) a fingerprint moved during acquisition; (c) an incorrectly placed fingerprint; and (d) a fingerprint acquired with too much pressure. (R. Stewart, M. Estevao, and A. Adler, Fingerprint recognition performance in rugged outdoors and cold weather conditions, in *Proceedings of the 3rd IEEE International Conference on Bio-metrics: Theory, Applications, and Systems*, pp. 1–6. © 2009, IEEE.)**

method described in [220] combines local and global characteristics and is based on features related to the effective area, energy concentration, spatial consistency, and directional contrast.

One of the most commonly used quality estimation techniques in the literature is described in [221]. This technique is specifically designed for use in fingerprint recognition systems based on identity comparison algorithms that use minutiae features, such as the algorithm presented in [186]. This method is based on neural networks and classifies five levels of quality, from poor to excellent. The features used are computed using the local quality map and the quality of the minutiae points estimated using the software MINDTCT [186]. The local quality map is computed considering the ridge orientation map and detecting regions with low contrast, low ridge flow, and high curvature. The quality of every minutia is obtained by considering its coordinates and by statistical analysis (mean and standard deviation) of the intensity values of the local image regions centered in the minutiae coordinates.

4.5.3 Image Enhancement

Another potential step in the fingerprint recognition process is image enhancement. Different enhancement techniques are available for fingerprint images [5]; these techniques can be classified as pixel-wise enhancement, contextual filtering, and multiresolution enhancement.

Pixel-wise enhancement techniques are based on image processing algorithms for enhancing image intensity [182]. These techniques are usually adopted as a preprocessing task during fingerprint image enhancement. Examples of these techniques are Wiener filtering [222] and the intensity-normalization algorithm proposed in [168]. The algorithm described in [168] is one of the most commonly used preprocessing techniques and is based on the following formula:

$$I'(x,y) = \begin{cases} \mu_0 + \sqrt{\dfrac{(I(x,y) - \mu)\sigma_0}{\sigma}} & \text{if } I(x,y) > \mu, \\[3ex] \mu_0 - \sqrt{\dfrac{(I(x,y) - \mu)\sigma_0}{\sigma}} & \text{otherwise,} \end{cases} \tag{4.2}$$

where μ_0 and σ_0 are the desired mean and standard deviation, respectively, of the normalized image I', and μ and σ are the mean and standard deviation, respectively, of image I.

The contextual filtering techniques are the most commonly used in the literature. These techniques change the characteristics of the filter used for enhancing different fingerprint regions according to the local context. The filters used perform an averaging effect along the ridges to reduce the noise and to enhance the contrast between ridges and valleys by performing band-pass filtering according to the ridge orientation. The most commonly used contextual filtering technique is based on Gabor filters tuned according to the local ridge characteristics [168]. This method computes two images that describe the local orientation O_R and the local frequency F_R of the ridges and then applies a convolution to the local areas of the original images with Gabor filters tuned according to the frequency and orientation of the ridges. The schema of the enhancement process is presented in Figure 4.13. The

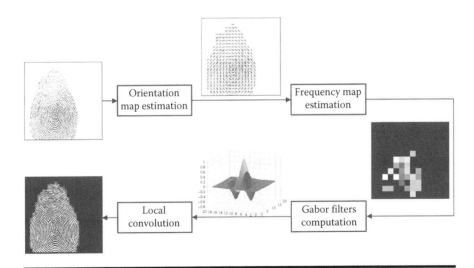

Figure 4.13 **Schema of a fingerprint enhancement method based on a contextual filtering technique.**

even-symmetric Gabor filter has the following general form:

$$h(x, y : \phi, f) = \exp\left\{-\frac{1}{2}\left[\frac{x_\phi^2}{\sigma_x^2} + \frac{y_\phi^2}{\sigma_y^2}\right]\right\}\cos(2\pi f x_\phi), \tag{4.3}$$

$$x_\phi = x\cos(\phi) + y\sin(\phi), \tag{4.4}$$

$$y_\phi = -x\cos(\phi) + y\sin(\phi), \tag{4.5}$$

where ϕ is the orientation of the Gabor filter, f is the frequency of a sinusoidal plane wave, and σ_x and σ_y are the space constants of the Gaussian envelope along the x- and y-axes, respectively. The parameters ϕ and f of the Gabor filter applied to the pixel (x, y) of the image I are selected according to $O_R(x, y)$ and $F_R(x, y)$, respectively. The parameters σ_x and σ_y are empirically tuned.

Other contextual filters are based on different masks. For example, the method described in [223] is based on bell-shaped filters, and the method proposed in [224] is based on log-Gabor filters. Other methods perform similar tasks, such as the method described in [164], which enhances local regions in the Fourier domain according to the maps describing the ridge orientation and frequency, and the method described in [225], which filters the fingerprint image in the frequency domain first and then applies local directional filters in the spatial domain.

Enhancement methods based on multiresolution techniques perform analyses at different scales (in the space domain or in the frequency domain) to first enhance the ridge structure, followed by finer details. An example of these methods is described in [226] and is based on a Laplacian-like image-scale pyramid.

Techniques for detecting and compensating for the presence of creases in fingerprint images have also been reported [227]. Creases can produce artifacts in enhanced images.

4.5.4 Feature Extraction and Matching

Fingerprint-matching algorithms compute a similarity index called a matching score between two fingerprints used in the verification/identification procedures. The fingerprint-matching algorithms can be divided into three different classes: correlation-based techniques, minutiae-based methods, and other feature-based methods.

4.5.4.1 Correlation-Based Techniques

Correlation-based techniques compute the matching score between two fingerprint images. The images are usually scaled, translated, rotated, and equalized, and then the matching score is obtained as a correlation measurement between the two images. The basic technique consists of computing the cross-correlation values [182]. Other methods are based on correlation techniques more robust to the presence of noise [228], and some methods can perform correlations between local regions [229]. The correlation-based approach is rather basic and not commonly employed in automated fingerprint recognition systems. This approach is not robust to distortions and intensity differences due to different finger pressures on the sensor.

4.5.4.2 Minutiae-Based Methods

Minutiae-based methods are the most studied and applied in the literature [230].

These methods compute the matching score between two templates T and T' by searching the corresponding minutiae in the considered feature sets. The minutia type is rarely considered because noise and different finger pressures on the sensor can easily transform a bifurcation into a termination and vice versa. Two minutiae correspond if their spatial distance s_d and direction difference d_d are less than the fixed thresholds. The distances s_d and d_d are defined as follows:

$$s_d(m'_j, m_i) = \sqrt{(x'_j - x_i)^2 + (y'_j - y_i)^2},$$
$$d_d(m'_j, m_i) = \min(|\theta'_j - \theta_i|, (360 - |\theta'_j - \theta_i|)). \tag{4.6}$$

Template registration must be performed to evaluate the number of corresponding minutiae points. Registration is essentially a point pattern-matching problem that aims to compensate for the rotations and translations of the templates to align the minutiae sets. Regardless, the algorithms used should be robust to different non-idealities, including the following:

■ The presence of false minutiae in the templates.
■ Missed minutiae.
■ Distortions due to the placement of the finger on the sensor.
■ Differences in the number of minutiae appertaining to the templates T and T'.

After the registration step, the corresponding minutiae are searched and used to compute the matching score. Different techniques to compute the matching score between two minutiae sets have been reported. One of the most diffused matching score formulas is presented in [5] as follows:

$$\text{Matching score} = \frac{k}{(M + N)/2}, \tag{4.7}$$

where k is the number of matched minutiae pairs, and M and N are the numbers of minutiae in the templates T and T', respectively.

Considering the used registration strategy and technique for searching the corresponding minutiae pairs, global and local minutiae-matching algorithms can be distinguished. Global algorithms use all of the minutiae points of the two finger-print templates and search the best matching score by aligning the two minutiae sets. Local algorithms consider sets of minutiae divided into subportions, for example, by adopting auxiliary graph structures or additional information related to the local regions of the minutiae points. Some algorithms are also specifically designed to compensate for problems related to skin distortions.

Many different global minutiae-matching algorithms have been reported. These algorithms can be based on algebraic geometry [231], Hough transform [232], relaxation [233], and energy minimization [234], among others. Some methods perform a prealignment of the query template with the corresponding template stored in the database to reduce the time required by the matching step [235].

Methods based on algebraic algorithms are the most diffused and rectify the considered templates using heuristics based on geometrical equations. The well-known software NIST BOZORTH3 [186] is appropriate for this class. The matching method can be divided into the following three steps:

1. Construction of intrafingerprint minutiae comparison tables: A table for template T and a table for template T' are computed. These tables describe a set of measurements of each minutia in a fingerprint relative to all other minutiae in the same fingerprint.
2. Construction of an inter-fingerprint compatibility table: The minutiae stored in the previously computed tables are compared, and the obtained results are stored in a distinct table.
3. Traversal of the inter-fingerprint compatibility table: The associations stored in the inter-fingerprint compatibility table represent single links in a compatibility graph. An iterative algorithm traverses the compatibility graph to identify the longest path of linked compatibility associations. The matching score is considered the length of the longest path.

Local minutiae-matching methods are based on comparisons between local characteristics of the fingerprint images that are invariant to rotations and to translations. These characteristics can be used to directly compute a matching score value or to perform the registration between the considered templates. Many matching algorithms based on local characteristics have been reported. Methods that use nearest-neighbor-based structures, fixed radius-based structures, minutiae triangles, and texture-based local structures can be distinguished. A well-known method that uses a nearest-neighbor-based structure is presented in [236]; this method performs a local comparison by evaluating the characteristics (coordinates and orientation) of the nearest l minutiae. Methods that use fixed-radius-based structures extract local features by considering local regions centered in every minutiae point. An example of these techniques is proposed in [237]. For each minutia, this method computes a graph that represents a star, which is obtained by connecting the considered minutia and the n nearest minutiae present in the evaluated local region. Methods based on minutiae triangles compute graphs that describe the connections between the nearest minutiae points. Many recognition algorithms in the literature use this type of template to perform the template registration because using minutia triangles permits accurate results to be obtained and reduces the computational time. One of the most commonly used techniques for computing these graphs is the Delaunay triangulation [238–240]. An example of the results of the Delaunay triangulation is presented in Figure 4.14.

Techniques based on the graphs obtained by applying the Delaunay triangulation search the triangles corresponding to two templates by evaluating characteristics that are invariant to rotations and translations. Examples of these characteristics are the angles, lengths of the facets, and lengths of the bisectors of the triangles. The most important problem with these methods is that these methods are sensitive to false and missed minutiae.

Other techniques in the literature have been specifically designed to compensate for nonlinear distortions due to differences in finger pressure on the sensor. A matching method based on a specifically designed feature, referred to as the local relative location error descriptor, is proposed in [241]. Other methods are based on

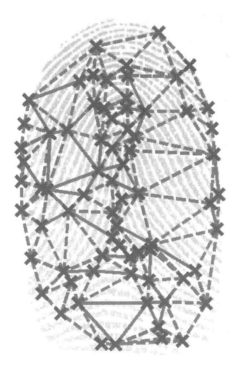

Figure 4.14 Visual example of the template used by local minutiae-matching methods based on graphs obtained by the Delaunay triangulation of minutiae.

multiple registrations [242], algorithms that estimate the mean distortion present in the considered data [243], and techniques for estimating the distortion models obtained by different finger placements on the sensor [244].

4.5.4.3 Other Feature-Based Methods

Other fingerprint matching methods use features extracted at different levels as support information for processing a matching score based on the minutiae sets, or directly process features extracted at Level 1 or at Level 3.

Most of these matching methods are based on texture analysis performed at Level 1. One of the most commonly used techniques in the literature is based on the FingerCode template [180]. Figure 4.15 presents the schema of the method based on the FingerCode template. This method can be divided into the following steps:

- Estimation of the core point.
- Definition of the region of interest (ROI) as a ring with a fixed size (height H).
- Partition of the ROI in N_R rings and N_A arcs to obtain $N_S = N_R \times N_A$ sectors S_i.
- Application of a bank of N_F Gabor filters with different directions to the image to obtain N_F filtered images $F_{i\theta}(x, y)$.

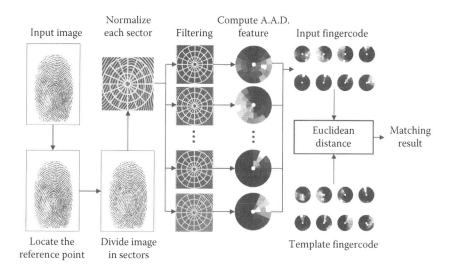

Figure 4.15 Schema of the biometric recognition method based on the template Fingercode. (M. Barni et al., A privacy-compliant fingerprint recognition system based on homomorphic encryption and fingercode templates, in *Proceedings of the 4th IEEE International Conference on Biometrics: Theory Applications and Systems*, pp. 1–7. © 2010, IEEE.)

■ Computation of the average absolute deviation (AAD) from the mean of gray values in individual sectors of filtered images to define the feature vector that represents the biometric template. The value $V_{i\theta}$ of the template related to every sector of each filtered image is computed as follows:

$$V_{i\theta} = \frac{1}{n_i}\left(\sum_{1}^{n_i}|F_{i\theta}(x,y) - F_{i\theta}|\right), \qquad (4.8)$$

where n_i is the number of pixels in S_i and $P_{i\theta}$ is the mean of the pixel values $P_{i\theta}$ of $F_{i\theta}(x,y)$ in sector S_i.

The obtained feature vector is composed of $N_V = N_S \times N_F$ values (e.g., in [180], N_V ranges from 640 to 896 depending on the fingerprint dataset used). This method is not rotation-invariant. Thus, during the enrollment phase, N_θ templates related to different rotations of the original image are computed. The matching score from two templates consists of the minimum Euclidean distance between the N_θ rotated templates and the live template. This step reduces problems originating from different finger placements on the sensor.

Another well-known method is described in [245] and is based on templates obtained from analyzing the circular regions of the image spectrum. Other matching methods are based on different local and global characteristics, for example, the method described in [246] compares scale-invariant features (SIFT) obtained from two fingerprint samples.

Some matching methods in the literature are also based on a Level 1 analysis that computes features describing the shape of every ridge [247].

Some methods are based on Level 3 features. The recognition method described in [204] extracts the pore features using Gabor filters and wavelet transforms and performs local matching using the iterative closest point (ICP) algorithm. Other matching methods based on Level 3 characteristics are described in [248,249].

4.5.5 Fingerprint Classification and Indexing

During identification, biometric systems compare the fresh biometric acquisition with all of the templates stored in a database. In some application contexts, the required computational time can be unacceptable. One useful strategy reduces the number of templates compared by partitioning the database into bins containing only data pertaining to a defined class. Many applications use the Galton–Henry scheme, which performs a classification in five classes by evaluating Level 1 features. The following classes are considered: the arch, tented arch, left loop, right loop, and whorl. Examples of images pertaining to these classes are presented in Figure 4.16. Methods for performing fingerprint classification [250] include the following: neural network classifiers, statistical methods, syntactic methods, rule-based methods, and support vector machines (SVMs). To further reduce the number of identity comparisons, more complex strategies could be adopted. Some methods

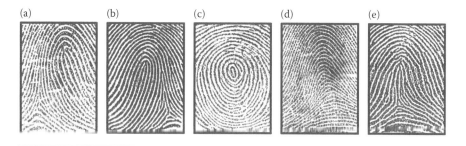

Figure 4.16 Galton–Henry classification scheme: (a) left loop, (b) right loop, (c) whorl, (d) arch, and (e) tented arch.

perform subclassification [251], and some strategies compute continuous indices from different fingerprint features [240].

4.5.6 Computation of Synthetic Fingerprint Images

Large datasets of biometric samples are required to design and evaluate fingerprint recognition systems. However, creating these datasets is expensive and time-consuming. Moreover, acquisition sensors may be modified during this process, increasing the efforts required to collect biometric data. In this context, the use of techniques for simulating synthetic samples can assist the acquisition of large sets of biometric data with limited resources.

The technique presented in [252] is based on computing global and local features. First, global features are computed to obtain a map of the ridge orientation. Then, local features are synthesized to simulate minutiae points. This step is performed by first defining a set of points with a binary mask, followed by iterative refinement of the set using a striped filter.

A biologically inspired mathematical model for simulating ridge pattern formation is proposed in [253,254]. This approach is based on the idea that the fingertip is created by the differential growth of the basal skin layer. Synthetic fingerprint images are computed by simulating the compression that creates the ridge pattern.

A genetic algorithm is used in [255]. This method uses a genetic algorithm to adapt a set of filters applied to real fingerprint images to obtain synthetic samples. This approach permits the process to be adapted for computing synthetic samples similar to the real fingerprints of a defined set of users.

The technique presented in [256] consists of several steps. First, the fingerprint area is defined using a silhouette. Then, the ridge orientation map is computed starting from the coordinates of the singular points (loops and deltas). A heuristic approach is applied to obtain the ridge frequency. Then, the ridge pattern is computed using a contextual iterative algorithm based on Gabor filters. Finally, the realism of the fingerprint image is improved by simulating the displacement, rotation, distortion, skin condition, and acquisition noise.

The method described in [257] improves the technique presented in [256] by enabling the simulation of fingerprints of individuals with characteristics similar to those of the final users of the biometric system to be tested (such as gender, age, and race). Thus, real biometric applications can be better simulated.

The study in [258] aims to increase the realism of synthetic fingerprint images by simulating texture-characterizing features estimated from real data, such as the ridge intensity along the ridge centerlines, ridge width, ridge cross-sectional slope, ridge noise, and valley noise.

4.6 Touchless Fingerprint Biometrics

Most fingerprint recognition systems use touch-based acquisition devices. However, these systems suffer from the following important intrinsic problems [259]:

- *Inconsistent contact*: The contact of the finger with the acquisition sensor causes distortions in the captured images due to elastic deformations of the friction skin of the finger. The introduced deformations can have different magnitudes and directions for each acquisition.
- *Nonuniform contact*: Different factors can introduce noise and reduce the contrast of the local regions of fingerprint images, including the dryness of the skin, shallow/worn-out ridges (due to aging or genetics), skin disease, sweat, dirt, and humidity in the air.
- *Latent print*: Each time a user places a finger on the sensor platen, a latent fingerprint is left, posing a security problem for this system because latent fingerprints can be used to perform impostor accesses. Moreover, during the acquisition process, the device can capture both the new fingerprint and portions of latent fingerprints, resulting in inconsistent samples.

Touchless fingerprint recognition systems are being studied to overcome these problems. In fact, touchless biometric systems permit biometric samples to be obtained that lack distortions originating from the contact of the finger with the sensor, that are more robust to dirt and to different environmental conditions, and that do not present latent fingerprints on the sensor surface.

Another important goal of touchless fingerprint recognition systems is to increase user acceptability compared with touch-based techniques. In fact, cultural factors and fears related to the transmission of skin diseases can limit the acceptability of touch-based recognition systems. Moreover, the use of touchless acquisition techniques can reduce the efforts necessary for user training and the time required for each biometric acquisition.

Touchless recognition systems are typically based on images captured by CCD cameras. These images are extremely different from those obtained using touch-based acquisition sensors. Examples of fingerprint images captured using a CCD camera and a touch-based sensor are presented in Figure 4.17. The touchless

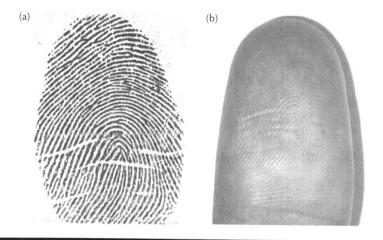

Figure 4.17 **Examples of fingerprint images captured using a touch-based sensor and a CCD camera: (a) touch-based image and (b) touchless image. Touchless fingerprint images are noisier and present backgrounds that are more complex.**

fingerprint images clearly feature more noise, reflections, and a more complex background than touch-based images. Moreover, the skin can be considered part of the background. Other problems that should be considered when designing touchless recognition systems are variable resolutions and differences in the acquisition angle.

Most fingerprint recognition systems in the literature can be characterized by three primary steps: acquisition, computation of a touch-equivalent fingerprint image, and feature extraction and matching. The schema of the biometric recognition process is presented in Figure 4.18.

Different touchless acquisition strategies [260] based on single cameras, multiple-view techniques, structured light approaches, photometric stereo systems, depth from focus methods, or acoustic imaging have been described in the literature.

The methods for computing touch-equivalent fingerprint images are strictly dependent on the acquisition setup employed. The goal of these methods is to obtain fingerprint images that can be used by algorithms designed for touch-based recognition systems. For systems based on single touchless fingerprint images, touch-equivalent fingerprint images are obtained by applying specifically designed enhancement techniques and by normalizing the image size to a standard resolution (usually 500 ppi). Other systems compute a metric reconstruction of the fingertip and then apply techniques for mapping three-dimensional data into a two-dimensional space. The computation of three-dimensional models permits less-distorted samples to be obtained but requires complex and expensive acquisition setups.

Finally, feature extraction and matching techniques are applied to perform biometric recognition. Most of the systems in the literature use methods designed for touch-based fingerprint images; however, algorithms specifically designed for

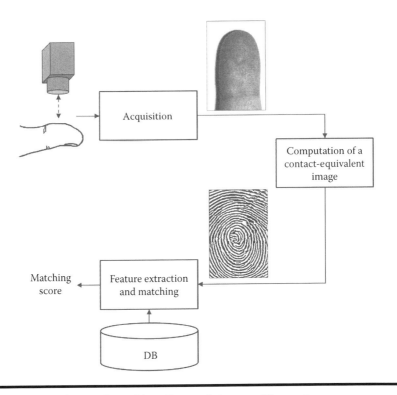

Figure 4.18   Schema of touchless fingerprint recognition systems.

touch-equivalent fingerprint images and identity comparison strategies based on three-dimensional features have also been explored.

This section presents reported techniques for acquiring and processing touchless fingerprint samples. The described techniques are divided into two classes: methods based on two-dimensional and three-dimensional samples.

4.6.1 Fingerprint Recognition Based on Touchless Two-Dimensional Samples

Reported touchless fingerprint recognition systems include those based on two-dimensional systems designed to be integrated in low-cost and portable devices and those employing more complex hardware setups to obtain higher recognition accuracy.

Touchless fingerprint recognition systems based on two-dimensional samples use CCD cameras to capture the details of the ridge pattern. These systems usually capture a single image and then process this image to obtain a touch-equivalent fingerprint image. This task aims to compute a fingerprint image compatible with existing biometric recognition systems designed for touch-based images. Then, biometric recognition is performed using well-known algorithms for extracting features and matching touch-based fingerprint images. Other systems use feature extraction and matching algorithms specifically designed for touchless two-dimensional samples. Some systems also use multiple-view techniques or specifically designed optics to obtain touch-equivalent fingerprint images not affected by perspective deformations and focusing problems.

4.6.1.1 Acquisition

The most important factors that characterize the acquisition setups used to capture touchless finger images are the use of supports for finger placement, the adopted optical configuration, and the illumination technique. In fact, the obtained images can present different characteristics and levels of noise. As an example, Figure 4.19 presents some images obtained using the same point light source placed in different positions.

The simplest acquisition technique uses a low-cost CCD camera under uncontrolled light conditions. A biometric system that captures fingerprint images using a webcam under natural light conditions is presented in [262], a study regarding images captured using a digital single-lens reflex camera under uncontrolled illumination conditions is proposed in [263], and studies regarding the use of mobile phone cameras are described in [264,265]. However, images captured under uncontrolled light conditions present poor contrast between ridges and valleys. Thus, most reported touchless recognition systems use illumination techniques to improve the visibility of the fingerprint. A simple and low-cost illumination technique uses a point light source, such as a lamp [266–268]. The primary disadvantage of point light sources is the introduction of shadows that can reduce the visibility of the ridge pattern. Moreover, such light sources do not permit uniform illumination of

(a) (b) (c) (d)

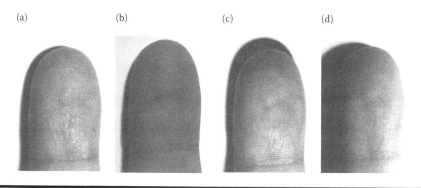

Figure 4.19 Touchless fingerprint images captured under different illumination conditions. (a) frontal illumination, (b) illumination from the left side, (c) illumination from the right side, and (d) illumination from the top side. (R. Donida Labati et al., Virtual environment for 3-D synthetic fingerprints, in *Proceedings of the IEEE International Conference on Virtual Environments, Human-Computer Interfaces and Measurement Systems*, pp. 48–53. © 2012, IEEE.)

the external regions of the fingerprint images. Other methods use ring illuminators [269] to overcome these problems and to obtain uniform illumination in all regions of the images.

Studies have also examined light wavelengths that enhance the visibility of the ridge pattern [259,270]. These studies report that white light does not provide the best contrast between ridges and valleys. In fact, long-wavelength rays, such as white and infrared light, tend to penetrate the skin and be absorbed by the epidermis. In contrast, a blue light with a wavelength of 500 nm results in lower hemoglobin absorption and enhances the details of the skin. Thus, some touchless fingerprint acquisition systems use illumination techniques based on blue LED lamps [271,272]. The work presented in [270] compares illumination setups based on different light wavelengths, light positions, polarization, and diffusion techniques. The best quality images are obtained using a horizontally polarized blue light with a tilt angle of $45°$ that has been treated with a scattering filter to obtain uniform illumination of all regions of the finger.

Other acquisition systems use transmission-based illumination techniques. The system described in [273] uses a red light source placed on the fingerprint side to focus the light transmitted through the finger onto a CCD. This method captures images describing the shape of the ridges in the internal layers of the finger and is less sensitive to problems related to bad skin conditions than other illumination techniques. However, the acquisition setup requires the finger to be placed in a fixed position, imposing constraints on the user.

Another important aspect of touchless acquisition systems is that these systems must guarantee a proper depth of focus and field of view to capture the details of all regions of the fingertip. To obtain an appropriate field of view, most of the systems in the literature require that the distance between the finger and the camera be less than 10 cm. Considering the required magnification ratio and the cylindrical shape of the finger, standard lenses can obtain fingerprint images affected by out-of-focus problems in the lateral regions. To overcome these problems, some systems use multiple cameras [272,274,275]. Another possibility is the use of curved lenses. However, such lenses can produce distortions in the captured images and increase the costs of the hardware setup.

Other problems that should be addressed are related to image resolution and to motion blur. The use of supports for finger placement allows the acquisition setup to be calibrated to estimate the image resolution and to reduce the probability of finger movements. Thus, many systems use supports for finger placement [259,269,274–276]. Less-constrained setups that do not require the use of supports have also been examined [272].

However, touchless fingerprint images captured by single cameras do not feature constant resolution in the different fingerprint regions. In fact, in this type of acquisition system, the optical resolution decreases from the detector center to the detector side due to finger curvature. This factor can also be influenced by the small focus field of single cameras with high-magnification lenses.

Most reported acquisition systems allow fingerprint images to be acquired from a single finger. A device for acquiring biometric samples from five fingers is presented in [276]. This device requires the fingers to be placed in five holes of a cylindrical holder. A rotating camera acquires the finger images at different times. The illumination system consists of two arrays of green LEDs.

The size of the fingerprint image database in some governmental applications is extremely large. Thus, compression algorithms are applied to the biometric samples. The study in [277] evaluated the results obtained using different techniques to compress touchless fingerprint images. The WSQ technique [211] achieved the best performance.

4.6.1.2 Computation of a Touch-Equivalent Image

Samples captured by touchless sensors typically cannot be directly used by recognition methods designed for touch-based fingerprint images. For compatibility with these biometric algorithms and the existing AFIS, most reported touchless fingerprint recognition systems compute touch-equivalent fingerprint images that represent the ridge pattern at a fixed resolution.

The performances of commercial fingerprint recognition software with touchless images were evaluated in [265]. Sufficient results were only obtained with the best-quality images.

Ridge pattern enhancement is usually computed before performing biometric recognitions based on well-known methods in the literature. This task aims to obtain a gray-scale image representing only the fingerprint pattern and to reduce the noise present in the captured touchless image. An example of a touchless fingerprint image and its corresponding touch-equivalent image are presented in Figure 4.20.

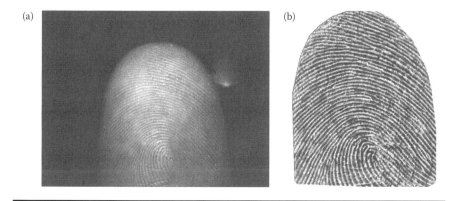

(a)   (b)

Figure 4.20 Example of results obtained by computing a touch-equivalent image from a touchless fingerprint image: (a) a contactless image and (b) its corresponding touch-equivalent image.

Important goals for computing touch-equivalent fingerprint images include increasing the contrast between ridges and valley by removing the details of the finger skin, reducing out-of-focus problems, removing possible reflections, and reducing the noise introduced by CCD cameras. Different techniques for enhancing touchless fingerprint images have been designed. The method described in [262] first performs a preprocessing task based on the Lucy–Richardson algorithm and on input image deconvolution using a Wiener filter. This task aims to reduce out-of-focus problems and to decrease the presence of noise. Then, a background subtraction algorithm based on low-pass Gabor filters is applied. Finally, a cut-off filter tuned according to the mean ridge frequency is used to increase the contrast between ridges and valleys. In contrast, the image enhancement method proposed in [266–268] is based on a contextual filtering technique. First, this method estimates the fingerprint area by converting the captured color image in gray-scale and applying a segmentation technique based on an adaptive threshold and morphological operators. Finally, the visibility of the touch-equivalent image is obtained using the contextual filtering technique based on the STFT analysis [164]. Another enhancement method designed for touchless fingerprint images is described in [278]. Similar to the technique presented in [168], this method applies Gabor filters tuned according to the local ridge frequency and orientation but computes the ridge orientation map using an iterative regression algorithm designed to be more robust to the noise present in touchless fingerprint images.

To obtain touch-equivalent fingerprint images that can be used by matching techniques based on minutiae features, the touchless images must also be normalized to a fixed resolution. Some touchless recognition systems that require the placement of the finger at a fixed position obtain this result by evaluating the information related to the focal length and the distance between the finger and the camera [259,269]. Systems that do not impose constraints can only perform an approximated normalization. The method presented in [262] first computes an alignment of the fingerprint image by applying a rotation inverse to the angle of the major axis of the finger silhouette. The final normalization task is registration of the minor axis of the image to the standard measure of 9/10 of the height of the final touch-equivalent image.

Preprocessing techniques for touchless images captured using mobile phones are reviewed in [279].

4.6.1.3 Touch-Equivalent Samples from Multiple Images

Touch-equivalent images obtained using single CCD cameras present problems related to perspective distortions and different resolutions in the local image regions. To overcome these problems, a two-dimensional touchless recognition system based on mosaicking fingerprint images obtained from three views is presented in [274]. The acquisition system is composed of a camera placed in front of the finger and two cold mirrors with a fixed tilt angle that are used to both obtain different views of the fingertip and function as a support for finger placement. The touch-equivalent

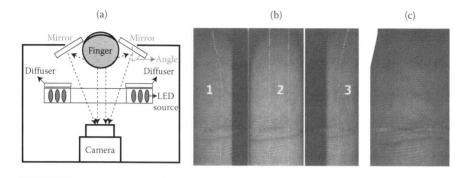

Figure 4.21 Multiple-view technique for computing two-dimensional fingerprint samples. (a) the acquisition setup, (b) examples of captured images, and (c) the corresponding mosaicked image. (H. Choi, K. Choi, and J. Kim, Mosaicing touchless and mirror-reflected fingerprint images, *IEEE Transactions on Information Forensics and Security*, vol. 5, no. 1, pp. 52–61. © 2010, IEEE.)

images are obtained by merging the information related to the different views. The schema of the acquisition setup is depicted in Figure 4.21a, an example of captured images is presented in Figure 4.21b, and the corresponding mosaicked image is presented in Figure 4.21c. The first task is image enhancement to increase the contrast between ridges and valleys in all captured images. The enhancement method used by this biometric system is similar to that presented in [278]; however, this method performs an adaptive histogram equalization task before applying the contextual filtering. The second task is the search for corresponding points in the three fingerprint images. To reduce the number of candidate corresponding points, the images are first rectified using the information obtained from a previously performed calibration of the multiple-view setups. The first set of corresponding points in the frontal and lateral views is obtained by applying a minutiae-matching technique. Starting from the obtained set of corresponding minutiae, a ridge-following approach is used to search for additional corresponding points. The next step consists of merging the three views. The regions of the lateral views that should be merged with the central region of the frontal view are selected by applying an iterative distance minimization technique to the set of corresponding points. Then, the lateral views are transformed according to the results obtained using the distance minimization and merged with the central region of the frontal view. Finally, the obtained touch-equivalent image is refined by smoothing the intensity values in the transitions between the merged regions. This system effectively reduces problems related to perspective distortions but not those due to differences in the resolution of local fingerprint regions because this system does not compute a metric representation of the biometric samples, in contrast to touchless systems based on three-dimensional reconstruction techniques.

A system for mosaicking images captured at different viewpoints is also described in [275]. The hardware setup consists of three cameras and four blue LEDs positioned around the finger placement guide. In contrast to the technique adopted in [274], the corresponding points between the different views are estimated using SIFT features and the RANSAC algorithm. Finally, the three-view images are fused in a mosaicked fingerprint image.

Another interesting system that considers the three-dimensional fingerprint shape to obtain a touch-equivalent fingerprint image is described in [280]. The acquisition setup comprises a beveled ring mirror and a CCD camera. The acquisition process requires the user to move a finger into the ring. During this process, the camera placed in front of the mirror captures a set of images that represent overlapping portions of the fingerprint. Then, the captured circular regions are mapped in rectangular areas using a log-polar transformation. The obtained images are merged using a correlation approach, producing an image that describes the complete fingerprint area. Finally, the visibility of the ridge pattern is enhanced. A less-expensive setup for this acquisition technique is also proposed in [280] and is based on a three-view configuration obtained using a line camera and two mirrors. The primary problem with this technique is that the speed and direction of the finger movement cannot be controlled and can introduce artifacts. Moreover, this system does not allow touch-equivalent images of constant resolution to be obtained for all fingerprint regions because this system assumes that the finger has a fixed shape.

4.6.1.4 Feature Extraction and Matching

Similar to touch-based fingerprint recognition systems, most biometric technologies based on touchless fingerprint images perform the recognition task using methods based on minutiae features because these methods can obtain accurate results and require limited computational time. These systems are usually based on matching techniques designed for touch-based images [259,262,274].

Matching algorithms based on minutiae features require constant image resolution to compute metric distances between the minutiae points. Thus, these algorithms can only be used if a proper resolution normalization is performed during touch-equivalent image computation. This task can be performed with sufficient accuracy only by adopting acquisition setups that use guides for finger placement because systems that impose fewer constraints on finger placement can only infer the image resolution from the characteristics of the image itself. In addition, adopting matching methods based on dimensional features is useful for performing recognition with sufficient accuracy in systems based on this type of acquisition setup. The matching technique proposed in [281] is specifically designed for touchless biometric systems that do not impose strong constraints on the placement of the finger during the acquisition task.

This technique is based on a feature set similar to the FingerCode template [180]. This method compares fingerprint templates using computational intelligence techniques to overcome problems related to perspective distortions and noise. Principal component analysis (PCA) is used to identify the most distinctive features, and SVMs are adopted to perform the template comparison.

A matching method specifically designed for low-resolution touchless images (approximately 50 ppi) is presented in [282]. This method can use features computed by applying Gabor filters with different orientations and features based on the localized Radon transform (LRT). Gabor filters are also applied to low-resolution images in [283] to extract information from both the finger skin and the finger vein pattern.

Multibiometric systems in the literature fuse the information from the analysis of the ridge pattern of touchless images and the corresponding vein pattern. The vein pattern can also be used to confirm the vitality of the finger in touchless acquisitions. Touchless fingerprint acquisition systems that permit images of the vein pattern to be captured are described in [272,284].

4.6.2 Fingerprint Recognition Based on Touchless Three-Dimensional Samples

Compared with biometric systems based on two-dimensional samples, systems that compute three-dimensional fingerprint models use more information and less distorted data. In fact, the samples used consist of three-dimensional structures that are not affected by perspective deformations and that represent a metric reconstruction of the fingertip. Moreover, the feature extraction and matching algorithms can use additional information related to the z-axis to improve the recognition accuracy. However, these systems require acquisition setups that are more complex and expensive than those for single touchless images. Moreover, most reported methods require complex acquisition procedures.

Three-dimensional biometric samples can be acquired using different methods. This step requires specifically designed hardware setups and three-dimensional reconstruction algorithms. The three-dimensional fingerprint reconstruction techniques described in the literature compute different types of samples. Some systems are able to estimate the three-dimensional shape of ridges and valleys [280,285–287], and some systems can represent the fingerprint as the three-dimensional volume of the finger with a superimposed texture that describes the ridge pattern [288–293]. Examples of a sample of the finger volume with the texture of the ridge pattern and a portion of a three-dimensional model representing ridges and valleys are presented in Figure 4.22.

Three-dimensional fingerprint models can be used directly by specifically designed feature extraction and matching techniques [294] or converted into touch-equivalent fingerprint images [265,288,295]. Dedicated techniques for directly comparing three-dimensional samples use additional information related to the three-dimensional coordinates of minutiae points in order to obtain recognitions that are more accurate. In contrast, touch-equivalent images are compared using existing recognition techniques designed for touch-based fingerprint images. Moreover, fingerprint images compatible with the existing AFIS can be obtained when proper strategies are used.

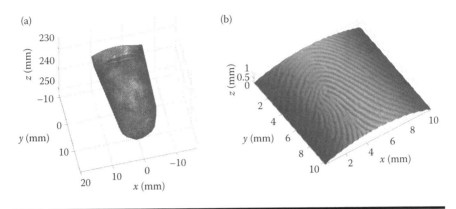

Figure 4.22 Examples of three-dimensional samples obtained using different techniques: (a) finger volume with the texture of the ridge pattern and (b) a portion of a three-dimensional model of ridges and valleys.

Most reported three-dimensional fingerprint recognition systems map the three-dimensional samples into a two-dimensional space to obtain touch-equivalent images. This task is usually called unwrapping or unrolling.

Three-dimensional fingerprint models have also been used for purposes other than biometric recognition. The approach described in [298] consists of a technique for printing three-dimensional fingerprint models. The resulting plastic models can be used to improve the accuracy of antispoofing techniques and the performance of both touchless and touch-based recognition systems.

4.6.2.1 Acquisition

Methods for computing three-dimensional fingerprint models are based on technologies including multiple-view techniques, structured-light approaches, photometric stereo systems, depth-from-focus methods, or acoustic imaging. Different technologies present advantages and disadvantages [299]. Methods for computing three-dimensional fingerprint models can also compute three-dimensional samples consisting of the finger volume with a superimposed texture representing the ridge pattern or models representing the three-dimensional shape of ridges and valleys. Methods that compute three-dimensional models representing only the finger shape usually perform faster acquisitions using less-expensive hardware setups. In contrast, methods that reconstruct the three-dimensional shape of the ridge pattern can use additional information in the feature extraction and matching steps of the biometric recognition process.

■ *Multiple-view techniques*. Three-dimensional reconstruction techniques based on multiple views first compute three-dimensional models by searching corresponding points in images acquired by cameras placed at different positions.

Then, the three-dimensional coordinates of these points are estimated using the triangulation algorithm. Reported methods based on multiple-view acquisition setups compute three-dimensional models representing the finger shape with a superimposed texture representing the ridge pattern. The primary advantage of these methods is that samples are acquired at a single time point.

A system for acquiring three-dimensional fingerprint models based on a multiple-view technique is presented in [280,285,300]. The acquisition setup uses five cameras arranged in a semicircle. These cameras point to the center of the semicircle, where the finger must be placed during the biometric acquisition. The illumination system consists of a set of green LEDs placed around the semicircle. Figure 4.23 presents the schema of the acquisition device. Three-dimensional reconstruction is performed using the information obtained from calibrating the multiple-view system, which is performed offline. The first step is a rough estimation of the finger volume and is performed using a shape from the silhouette technique. Then, the corresponding points in the images obtained from adjacent cameras are searched using a correlation-based technique. Finally, the three-dimensional shape is obtained using the triangulation algorithm [301], and the texture representing the ridge pattern

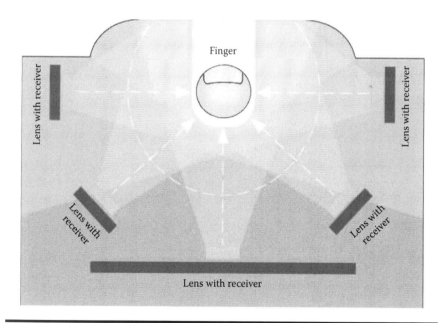

Figure 4.23 Schema of a multiple-view setup for the computation of three-dimensional fingerprint models. (Y. Chen et al., 3D touchless fingerprints: compatibility with legacy rolled images, in *Proceedings of the Biometrics Symposium: Special Session on Research at the Biometric Consortium Conference*, pp. 1–6. ©️ 2006, IEEE.)

is computed and superimposed on the three-dimensional model. Then, the model is composed of a depth map and a gray-scale image. The ridge pattern computation is based on an image enhancement method that can be divided into two distinct tasks. The first task is enhancing the ridge visibility, which is performed by applying a homomorphic filtering technique. First, the logarithm of the image is computed. Then, a high-pass filter is applied. Finally, the exponential of the obtained image is calculated. The second task consists of ridge pattern enhancement and is performed using the contextual filtering technique described in [168].

The method described in [286] uses an acquisition setup composed of three cameras and four blue LEDs. Corresponding points in the images are searched using the following characteristics of the fingerprint images: SIFT features, ridge map, and minutiae. Then, the RANSAC algorithm is applied to remove false correspondences. Next, the three-dimensional coordinates of the reference points are obtained by applying the triangulation algorithm. Finally, the finger model is computed by approximating the three-dimensional points to a previously defined shape.

Multiple-view methods have also been studied for the three-dimensional reconstruction of ancient fingerprints on artworks [303].

- *Structured-light approaches.* Structured-light scanners project sets of light patterns on a surface using illuminators and then infer the three-dimensional shape of the surface by analyzing the projected light patterns in different frames captured by a camera. Methods based on structured-light approaches can compute three-dimensional models of the ridge pattern or reconstruct the three-dimensional shape of the finger. A system able to estimate both the three-dimensional shape of ridges and valleys and the texture of the ridge pattern obtained from the visual aspect of the finger is presented in [288,289]. The acquisition setup of this system is presented in Figure 4.24 and comprises a camera and a projector. The three-dimensional reconstruction of the ridge pattern is obtained by projecting a sine-wave pattern shifted several times. A single frequency pattern is used with 10 phase-shift patterns. A frequency

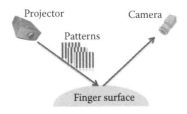

Figure 4.24 Schema of a structured-light setup for the computation of three-dimensional fingerprint models. (Y. Wang, L. Hassebrook, and D. Lau, Data acquisition and processing of 3-D fingerprints, *IEEE Transactions on Information Forensics and Security,* **vol. 5, no. 4, pp. 750–760. © 2010, IEEE.)**

of 16 cycles per length of each pattern is used. Each acquisition requires several frames to be captured. The three-dimensional shape is obtained by evaluating the phase shift of the projected pattern in every captured frame and by using the information related to a previous calibration of the system. Finally, a median filter is applied to the obtained depth map to remove possible spikes. The texture of the ridge pattern is computed by estimating the albedo image [304].

Another acquisition system based on a structured-light technique is described in [294]. In contrast to the device presented in [288,289], this system does not estimate the three-dimensional shape of ridges and valleys but only computes the finger volume and the texture describing the ridge pattern. The three-dimensional reconstruction technique used is based on the projection of a fringe pattern. The acquisition setup comprises a camera and a blue LED with a sinusoidal pattern. The finger volume is obtained by computing the phase map of every pixel and by using the calibration data to estimate the depth of every point with the triangulation algorithm [301]. The texture describing the ridge map is considered a touchless fingerprint image separately captured by the camera.

A study of the optimum configuration of the fringe pattern projected by a pico-projector for reconstructing three-dimensional fingerprint models [290] demonstrated that the best results were obtained using a 25-mm lens and by projecting a green fringe pattern perpendicular to the finger orientation. The time required to acquire the frames used to perform a three-dimensional reconstruction was 0.5 s.

■ *Photometric stereo systems.* Photometric stereo techniques estimate the surface normals of objects by analyzing images captured under different illumination conditions. Systems for acquiring the fingertip surface are based on sets of LEDs placed at known positions around the finger placement guide. The primary advantage of these systems is that they are usually based on less-expensive hardware setups compared to other techniques. However, accurate metric reconstructions of the finger shape are difficult to obtain using photometric stereo techniques.

The system presented in [294] is composed of seven LEDs and a single camera. The reconstruction method estimates the surface normals of the finger surface using seven images acquired by separately illuminating every LED and assumes that the reflectance properties of the skin can be approximated using the Lambertian model [305].

Tests of different hardware configurations and non-Lambertian illumination models are presented in [291,292].

■ *Depth from focus methods.* Three-dimensional models are obtained by analyzing sets of images captured by changing the focal length of the lens. Therefore, these three-dimensional reconstruction systems require autofocus mechanisms.

A method to reconstruct the three-dimensional shape of the finger is presented in [287]. To change the focal length of the lens, the acquisition setup uses an LCP (liquid crystal panel) and a birefringent optical positioned between the imaging element and the subject's finger.

■ *Acoustic imaging.* These systems are widely used in biomedical analyses because these systems permit the acquisition of images of internal organs.

A laboratory prototype of a four-transducer ultrasonic fingerprint scanner is presented in [293]. This acquisition method allows three-dimensional models describing the ridge pattern of the inner layer of the skin to be acquired and is robust to different environmental conditions. However, this method requires expensive hardware components.

4.6.2.2 Computation of a Touch-Equivalent Image

Computing touch-equivalent images based on three-dimensional fingerprint models involves mapping the three-dimensional shape into a two-dimensional space using unwrapping techniques. The goal of this task is to obtain data compatible with the existing AFIS and with recognition algorithms designed for touch-based fingerprint images. This task allows the acquisition of fingerprint images similar to those captured by inked acquisitions. An example of the obtained result is presented in Figure 4.25. The touch-equivalent image clearly does not present perspective distortions.

Different unwrapping techniques have been examined. These techniques can be divided into parametric methods and nonparametric methods. The first class of algorithms computes the projection of the three-dimensional fingerprint sample onto a parametric model (e.g., a cylindrical or a conic model) and then unwraps this model in a two-dimensional space. These methods are usually computationally efficient. However, differences between the approximating model and the real shape can introduce distortions in the unwrapped image. Nonparametric methods can be directly applied to the three-dimensional model without imposing constraints on its shape and can preserve local distances or angular relations. However, the mapping of a three-dimensional shape in a two-dimensional space does not permit the original metric distances between all points of the three-dimensional model to be maintained.

A simple parametric method is based on approximating the finger shape to a cylindrical model [265]. The first step estimates the center of rotation, which is considered the minimum z and mean x of the finger model. Then, the coordinates of every three-dimensional point are converted into cylindrical coordinates. Finally, the texture image of the ridge pattern is mapped in the new space using an interpolation technique. The primary problem with this method is that the obtained images present important horizontal distortions. Moreover, this method does not allow images with a fixed resolution in the horizontal axis to be obtained.

A method for reducing the distortion introduced by the approximation to a cylindrical model is proposed in [288]. First, the shape of the finger is approximated

(a)

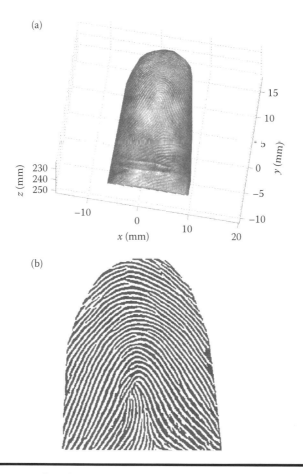

(b)

Figure 4.25 Example of a result obtained by unwrapping a fingerprint three-dimensional model: (a) a three-dimensional model and (b) its touch-equivalent image.

as a set of rings with different radii and center coordinates. Then, to reduce the noise present in the approximating model, a mobile averaging filter is applied to the estimated radii. Next, the approximating model is mapped into the two-dimensional space. If the three-dimensional map of the ridge pattern is available, then this method applies an iterative algorithm to obtain an image representing the three-dimensional ridge structure flattened on a plane. In this case, the best-quality regions of the images representing the ridge pattern obtained from the three-dimensional model and texture image are fused to produce the touch-equivalent image.

Another parametric method is described in [295]. The first step estimates a sphere that approximates the fingerprint model. Then, linear mapping into a two-dimensional space is performed, and a distortion correction algorithm based on the analysis of the distance between adjacent points is applied. The obtained result consists of a nonlinear map of the points pertaining to the three-dimensional model, which is used to compute the touch-equivalent image.

The nonparametric unwrapping technique described in [265] has been used by a variety of systems [280,285,306]. This method aims to maximize the preservation of the inter-point surface distances and scale. First, the fingerprint model is divided into slices along the vertical direction. Then, each slice is unfolded by a resampling algorithm that attempts to preserve the distances between the points pertaining to the slice. The sampling algorithm starts from the center of the slice and estimates the new (x, y) coordinates of every point according to the distance from the nearest points of the slice.

A nonparametric method based on mechanical lows is presented in [296]. The first step estimates the finger shape by removing the presence of ridges and valleys. This step is performed using a weighted linear least-square algorithm based on weights calculated by a Gaussian function. Then, the point cloud is considered a mechanical system in which each point is connected to 8-connected neighbors with virtual springs. An iterative algorithm searches the equilibrium position for each three-dimensional point and computes the touch-equivalent image based on the estimated positions.

However, three-dimensional fingerprint unwrapping methods usually obtain touch-equivalent images that are not completely compatible with existing databases of touch-based fingerprints. To increase the similarity between touch-equivalent images and touch-based images and, consequently, to increase the compatibility of touch-equivalent images with the existing AFIS, the method presented in [297] includes a simulation of the finger pressure on a touch-based sensor. This model assumes a higher pressure in the central region and a lower pressure in the lateral regions of the finger. The unwrapping method is based on the nonparametric technique described in [265]; however, this method determines a sampling interval for each point of the fingerprint according to its relative position with respect to the center of the ridge pattern. The obtained results are more compatible with existing touch-based biometric techniques than touch-equivalent images obtained by other methods.

In systems that compute three-dimensional models of the ridge pattern, specific algorithms for detecting ridges and valleys in three-dimensional matrices are also adopted [295,307]. These methods statistically evaluate local differences between the fingerprint model and approximating shapes.

4.6.2.3 Feature Extraction and Matching

Most of the studies in the literature compute contact-equivalent images and then apply feature extraction and matching algorithms designed for touch-based samples. Only a few studies have examined three-dimensional feature extraction and matching techniques based on three-dimensional characteristics.

In [294], a binary representation of the finger curvature (Finger Surface Code) is presented. The shape index [308] is computed and discretized in 15 classes. Each value is represented by a binary value of four bits. The matching algorithm computes the Hamming distance between Finger Surface Code templates.

A minutiae matcher based on three-dimensional coordinates is presented in [294]. This method represents every minutia using additional data compared to the standard two-dimensional representation. Two-dimensional minutiae are usually represented as (x, y, θ). In contrast, the method represents every minutia point as (x, y, z, θ, ϕ), where z is the minutia coordinate in the z-axis, and ϕ is the minutia orientation in spherical coordinates with unit length 1. Similar to traditional minutiae matchers, the presented algorithm estimates the matching score between two templates by iteratively aligning the minutiae templates and evaluating the number of corresponding minutiae pairs. Two minutiae are considered corresponding if the differences between their descriptors are lower than the fixed threshold values.

4.6.3 Applications of Touchless Fingerprint Biometrics

Market analysis performed by the International Biometric Group (IGB) [28] indicated that most biometric systems are based on the fingerprint trait and on touch-based acquisition techniques.

Only a few commercial touchless fingerprint recognition systems are currently available:

■ TBS 3D-Enroll and TBS 3D-Terminal from Touchless Biometric Systems [309], three-dimensional technologies based on multiple-view acquisition techniques.
■ TST Biometrics BiRD 3 [310], a three-dimensional system based on a multiple-view acquisition method.
■ Finger On the Fly from SAFRAN Morpho [311], a touchless sensor for the acquisition of four fingers.
■ Mitsubishi Finger Identification Device by Penetrated Light [312], a two-dimensional sensor that uses red illumination on the back of the finger.

Figure 4.26 presents examples of these technologies.

Companies are also studying algorithms to perform biometric recognitions based on touchless images captured under uncontrolled conditions [263].

Because touchless fingerprint technologies have been recently introduced, the diffusion of these technologies in real application scenarios has been limited. However, these technologies are appropriate for all applications that use live fingerprint acquisitions and have some advantages compared with touch-based recognition systems. For example, touchless systems could be applied in the government, commercial, and investigative sectors.

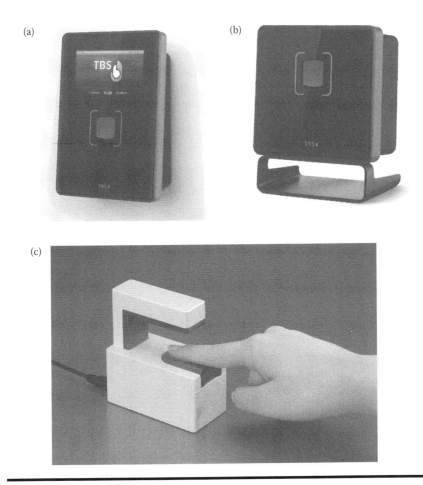

Figure 4.26 **Commercial sensors for the touchless acquisition of fingerprint samples: (a) TBS 3D TERMINAL (Reproduced with permission from Touchless Biometric Systems, TBS, http://www.tbs-biometrics.com), (b) TBS 3D ENROLL (Reproduced with permission from Touchless Biometric Systems, TBS, http://www.tbs-biometrics.com), and (c) Mitsubishi's Finger Identification Device By Penetrated Light. (Reproduced with permission from Mitsubishi Electric, http://www.mitsubishielectric.com).**

Border control is one important biometric application in the government sector. In this context, touchless acquisition techniques should reduce recognition time because of less-constrained biometric acquisition procedures. Biometric documents can also be created using touchless acquisition to reduce the costs of enrollment and verification devices.

Examples of applications in the commercial sector include authentication systems integrated into ATMs and terminal login. In this sector, user satisfaction is a critical aspect for the diffusion of these applications and services. Touchless fingerprint recognition systems should increase the user acceptability of biometric recognition process compared with touch-based systems.

Other applications have investigative uses. Touchless acquisition techniques can simplify the acquisition of samples from noncollaborative subjects compared with touch-based acquisition systems. However, the compatibility of the touch-equivalent images obtained by touchless systems with the existing AFIS is a crucial aspect for the diffusion of touchless technologies in this application context. No studies regarding this aspect have been performed using datasets with many samples.

Compared with other biometric technologies, touchless fingerprint recognition systems also increase the possibility of performing biometric authentications in web applications (e.g., e-banking, e-commerce, and e-government). In fact, recognition techniques based on single fingerprint images can obtain sufficiently accurate results using webcams or cameras integrated in mobile phones, without requiring dedicated biometric acquisition devices. Other biometric characteristics can be captured using the same devices (e.g., the face) but usually present less durability and accuracy.

The higher resolution of touchless images compared with touch-based fingerprint images should also permit the design of new applications, such as the network security protocol presented in [313].

4.7 Summary

Fingerprint biometrics is based on the analysis of the ridge pattern present on human fingers, a highly distinctive human characteristic. Moreover, the fingertip ridge structure is characterized by high durability, as this structure is fully formed at approximately the seventh month of fetal development and does not change throughout the lifetime of an individual unless serious accidents or diseases occur.

Fingerprint recognition systems are the most well-known and widely used biometric technologies. The primary application contexts of these systems are the forensic, government, and commercial sectors. Examples of applications in these sectors include investigative analyses, missing person searches, biometric documents, border control, terminal login, and authentication systems integrated into ATMs.

The analysis of fingerprint characteristics can be performed at three levels: global (Level 1), thin (Level 2), and ultrathin (Level 3). Level 1 evaluates the overall ridge flow, including characteristics such as the ridge orientation, local ridge frequency, singular regions, and ridge count. Level 2 analysis considers specific ridge discontinuities called minutiae. Many different classes of minutiae exist, but typically only bifurcations and terminations are considered. Many studies have examined methods for estimating minutiae points. The ultrathin characteristics evaluated at Level 3 are related to small details, such as pores and incipient ridges. This level of analysis can only be performed on high-resolution fingerprint images.

Fingerprint recognition systems typically perform different levels of analysis at each step of the recognition process. In traditional fingerprint biometrics, this process can be divided into the following steps: acquisition, quality evaluation, enhancement, feature extraction, and matching. The acquisition process can produce the following three different classes of images: latent fingerprints, inked fingerprints, and live-scan fingerprints. The quality of the captured image is then evaluated to discard samples that may produce erroneous recognitions. Next, the visibility of the ridge pattern is improved using enhancement techniques. The most commonly used methods in the literature can be classified into pixelwise enhancement, contextual filtering, and multiresolution enhancement. The next step is feature extraction, which is usually based on minutiae features. The final step is matching between biometric templates. In systems based on minutiae features, the matchers can consider local or global characteristics and can use techniques designed to overcome problems due to the presence of distortions in the fingerprint images. Matching methods based on correlation techniques and on Level 1 or Level 3 features also exist. Identification systems can also use a supplementary classification or indexing step to reduce the number of compared images. The most commonly used fingerprint classification schema is based on five classes (the arch, tented-arch, left loop, right loop, and whorl), which are estimated from the characteristics of the ridge orientation. Methods for computing synthetic fingerprint images, which can be used to design and evaluate new biometric algorithms, have also been reported.

Most fingerprint recognition systems use touch-based acquisition procedures. However, the obtained images can present nonlinear distortions and low-contrast regions due to variable finger pressures on the sensor platen. To overcome this problem and increase the usability and user acceptability of fingerprint biometrics, researchers are studying touchless recognition systems based on CCD cameras. However, touchless fingerprint images are extremely different from images captured by traditional sensors. Therefore, most touchless recognition systems aim to compute touch-equivalent fingerprint images, utilize traditional feature extraction and matching algorithms, and obtain data compatible with the existing biometric databases. Touchless fingerprint recognition systems can be based on two-dimensional and three-dimensional samples. Systems based on two-dimensional samples can use different acquisition setups based on one or more cameras and on dedicated illumination techniques. In these systems, touch-equivalent images are typically computed by applying image enhancement algorithms and resolution-normalization methods. However, the obtained touch-equivalent images usually present distortions due to perspective effects. Systems based on three-dimensional samples require acquisition setups that are more complex but can generally achieve results that are more accurate. Samples can consist of three-dimensional models of the ridge pattern or models representing the finger volume with a superimposed texture representing the ridge pattern. These three-dimensional models can be obtained using structured-light approaches or multiple-view techniques. To obtain touch-equivalent images, the three-dimensional models are mapped into a two-dimensional space by applying parametric or nonparametric unwrapping algorithms.

Touchless fingerprint recognition systems are appropriate for all application contexts in which traditional fingerprint biometrics are applied and feature great advantages in terms of usability and user acceptability.

Chapter 5

Touchless Fingerprint Recognition

This chapter presents methods for touchless fingerprint biometrics. An important characteristic of these approaches is the lack of finger placement guides, which reduces the acquisition constraints compared with most touchless fingerprint recognition systems described in the literature. To minimize the level of user cooperation required, all studied approaches are based on images captured in a single instant.

The researched approaches can be divided into methods based on two-dimensional and three-dimensional samples. The first class of techniques is designed for low-cost and portable applications. Although methods based on three-dimensional samples produce more accurate results, these methods utilize more complex hardware setups and algorithms.

The techniques described in this chapter include the following aspects of touchless biometric systems: sample acquisition and evaluation of biometric sample quality, computation of touch-equivalent images, matching algorithms, and computation of synthetic samples.

5.1 Touchless Fingerprint Recognition Techniques

Methods for all steps of the biometric recognition process based on two-dimensional and three-dimensional touchless fingerprint recognition systems have been studied. These methods are all designed to work under less-constrained conditions than those required by most previously reported touchless recognition technologies.

A schema of the researched biometric recognition techniques is presented in Figure 5.1. Four approaches for touchless fingerprint recognition systems have been studied: an approach based on two-dimensional samples, an approach based on three-dimensional samples and touch-equivalent images, an approach based on three-dimensional samples and three-dimensional templates, and a touchless approach based on three-dimensional minutiae points. Different techniques have been designed for all steps of the biometric recognition process based on touchless fingerprint images: (I) sample acquisition and evaluation of sample quality; (II) computation of three-dimensional samples; (III) computation of touch-equivalent fingerprint images; (IV) feature extraction; and (V) matching.

Different acquisition setups for recognition systems based on two-dimensional and three-dimensional samples have been studied. These hardware setups do not require the use of finger placement guides, are based on one or two cameras, and can capture the information required by the biometric recognition methods in a single instant.

A quality evaluation method able to estimate the best-quality frames in frame sequences describing a finger moving toward the CCD camera has been realized so that biometric acquisitions can be performed in uncontrolled applications.

Single touchless fingerprint images can then be used directly to compute touch-equivalent fingerprint images. The studied method for computing touch-equivalent images first performs an enhancement step to improve the visibility of the ridge pattern and to reduce the noise present in the touchless image, followed by normalization of the image to a constant resolution. Two algorithms for enhancing touchless fingerprint images have been implemented. The first algorithm is designed to function under particularly noisy conditions, whereas the second algorithm is more computationally efficient. The studied resolution normalization technique is designed for fingerprint images captured at a fixed distance from the CCD camera.

Because touchless fingerprint images present different non-idealities than touch-based samples, applying feature extraction and matching algorithms designed for touch-based samples to touchless fingerprint images can produce unsatisfactory results. Therefore, we have studied both Level 1 and Level 2 methods for analyzing touchless fingerprint images. The studied Level 2 methods consist of modifications of well-known techniques as well as a novel approach for compensating for perspective distortions and rotations of the finger. This approach is based on computational intelligence techniques and the computation of synthetic three-dimensional models of the finger shape. We have also designed a Level 1 analysis method for searching for the core point in touchless images, which uses computational intelligence techniques to search for the core in a list of candidate points.

Different methods for computing three-dimensional fingerprint models have also been studied, including a method that can estimate the three-dimensional coordinates of the minutia points and three different methods for computing three-dimensional models describing the complete finger surface. These methods are all based on multiple-view acquisition setups. The first method can only

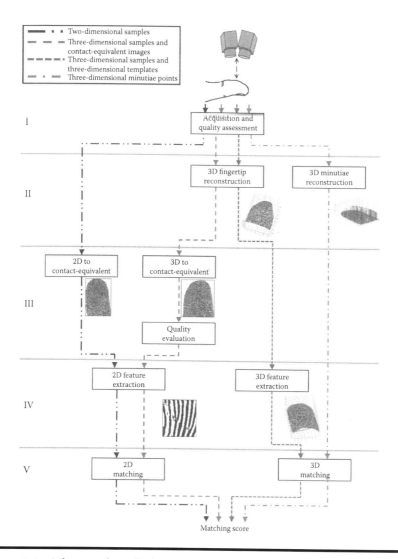

Figure 5.1 Schema of studied approaches for touchless fingerprint recognition, including approaches based on two-dimensional samples; three-dimensional samples and touch-equivalent images; three-dimensional samples and three-dimensional templates; and three-dimensional minutiae points. Techniques have been studied for all steps of the biometric recognition process based on touchless fingerprint images: (I) sample acquisition and evaluation of sample quality; (II) computation of three-dimensional samples; (III) computation of touch-equivalent fingerprint images; (IV) feature extraction; and (V) matching.

be used to perform biometric recognitions based on matchers that consider the three-dimensional feature points and is based on computational intelligence techniques. The methods for computing three-dimensional models describing the finger surface generate samples consisting of the three-dimensional fingerprint volume with a superimposed image representing the ridge pattern. These three-dimensional reconstruction techniques share a common computational schema: First, the ridge visibility is enhanced. Then, pairs of corresponding points are extracted from the images obtained from the different views. Next, a noise reduction technique is applied to the pairs of corresponding points. The three-dimensional finger volume is subsequently computed. Finally, the ridge pattern is estimated and superimposed onto the three-dimensional model.

Three-dimensional fingerprint model matching can be performed using algorithms that consider three-dimensional features or by applying unwrapping methods and then using techniques designed for touch-based fingerprint samples. Matchers based on three-dimensional minutiae points or unwrapping techniques have been investigated. The three-dimensional matcher compares Delaunay graphs computed from the three-dimensional coordinates of the minutiae points. In contrast, the unwrapping technique approximates the finger shape as a set of rings and then maps every ring to a two-dimensional space.

To improve the recognition accuracy by discarding insufficient-quality samples, a quality estimation method specifically designed for touch-equivalent images obtained from the unwrapping of three-dimensional models has also been developed. This method can detect non-idealities introduced by the three-dimensional reconstruction step and estimate a discrete quality value using computational intelligence classifiers.

The design and evaluation of biometric algorithms and acquisition setups require the use of large biometric databases. To reduce the efforts necessary to collect biometric samples, an approach for computing synthetic touchless fingerprint samples from touch-based fingerprint images has also been studied. This method can simulate three-dimensional fingerprint models obtained using different acquisition setups and illumination techniques.

This chapter is organized as follows: Section 5.2 presents the studied techniques for biometric systems based on touchless two-dimensional samples. This section describes the acquisition setup, quality estimation technique, and algorithms for estimating touch-equivalent images. Section 5.3 discusses techniques for three-dimensional model computation, three-dimensional matching, three-dimensional model unwrapping, and touch-equivalent image quality estimation for images obtained from unwrapping three-dimensional models. Finally, Section 5.4 presents the implemented approach for computing synthetic touchless fingerprint samples.

5.2 Methods Based on Two-Dimensional Samples

The researched approach for acquiring and elaborating touchless, two-dimensional fingerprint samples is presented in this section. First, the acquisition setup is

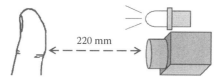

Figure 5.2 Schema of the proposed hardware setup for acquiring touchless fingerprint images.

presented. Subsequently, a quality assessment method for touchless images is described, and a technique for computing touch-equivalent images is discussed. Then, methods for analyzing Level 1 and Level 2 features in touch-equivalent fingerprint images obtained from touchless two-dimensional samples are described. Finally, a method for compensating perspective distortions and finger rotations is presented.

5.2.1 Acquisition

The proposed algorithms designed for touchless two-dimensional samples are based on fingerprint images captured using a single CCD camera. The acquisition setup is designed to be integrated into low-cost mobile applications. Thus, this setup does not require the use of finger placement guides. Another important aspect is the illumination technique. Complex and expensive illumination systems cannot be used to enhance the visibility of the ridge pattern under the described application conditions. Therefore, an LED light placed in front of the finger is used to enhance the visibility of the ridge pattern. Moreover, this setup can capture good-quality fingerprint images at a remarkable distance. All touchless fingerprint acquisition systems in the literature require the finger to be placed at a distance of less than 100 mm from the sensor. In contrast, the proposed acquisition setup captures touchless fingerprint images at a distance of more than 200 mm. Figure 5.2 presents a schema of the proposed acquisition technique.

5.2.2 Quality Assessment of Touchless Fingerprint Images

Touchless fingerprint acquisitions performed without finger placement guides and without dedicated illumination techniques can produce poor-quality images, potentially resulting in recognition errors. Moreover, of hundreds of frames, only a small number of images will be of sufficient quality for use in biometric recognition. Thus, to perform this type of biometric acquisitions in uncontrolled applications, quality assessment techniques must be adopted to discard images of insufficient quality. We have proposed a quality assessment approach that can identify the best-quality frames in frame sequences describing a finger moving toward the camera. This approach is presented in [314].

A quality assessment technique for touchless fingerprint images should consider a set of non-idealities and aspects that influence the quality of the captured samples. Different shades can be present on the fingertip due to the environmental light and the convex shape of the finger; these shades are considered noise effects. In addition, errors in the focus of the lenses and the relative movements of the subject in front of the camera can introduce blurring effects. The acquired images are also affected by the noise introduced by the acquisition sensor and electronic components. The application context and user ability also influence sample quality. In addition, finger movement can cause the following four primary non-idealities during each touchless biometric acquisition:

1. If the finger is too away from the camera, the sample may be unclear, and the details of the ridge pattern may be insufficiently visible.
2. If the finger is too close to the camera, the captured image may be affected by blur because the finger is not in the focus range.
3. If the finger is rotated with respect to the camera, perspective effects may be present in the sample, and some regions of the ridge pattern may not be acquired.
4. If the finger is moving faster than the exposure time of the camera, the sample may present motion-blur.

The approach aims to estimate the quality of every frame of sequence in real time, thus permitting the complete biometric system to select the best touchless image/images for use in performing the biometric recognition. Qualitative measurements can be expressed using continuous or discrete values. Continuous measures can be expressed in a defined range (e.g., from 0 to 1) or without fixed bounds. Discrete values can be used to represent classes of quality and are usually represented by a set of integer values. Similar to the NIST Fingerprint Image Quality (NFIQ) software [221], the proposed approach performs discrete measurements, thus obtaining easily understandable classifications of the image quality.

The proposed approach includes two distinct methods, which are outlined in Figure 5.3. For each captured frame, the first step of the two methods is the ROI estimation. Method QA performs an additional segmentation step, extracts a set of features, and then performs the quality classification using computational intelligence techniques. Method QB is based on the NIST Fingerprint Image Quality (NFIQ) software [221], which cannot be directly applied to touchless fingerprint images. Hence, Method QB computes the enhancement of the ridge pattern by first applying a specific band-pass filter. The used enhancement algorithm does not permit accurate touch-equivalent images to be obtained but does permit sufficient details to be obtained to perform a quality evaluation of touchless images in a computationally efficient manner. The use of computationally efficient algorithms is necessary for integrating a quality-estimation technique in real-time biometric acquisition systems.

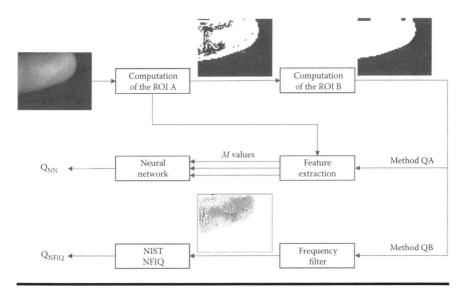

Figure 5.3 **Schema of the proposed quality-assessment approach. The left branch of the graph refers to Method QA, and the right branch of the graph refers to Method QB. (R. Donida Labati, V. Piuri, and F. Scotti, Neural-based quality measurement of fingerprint images in contactless biometric systems, in *Proceedings of the International Joint Conference on Neural Networks*, pp. 1–8. © 2010, IEEE.)**

5.2.2.1 Computation of the ROI

The input fingerprint for each captured frame is processed using a method based on the local variance of the input image for detecting the presence of the ridges in the ROI of the input image. A method based on the local variance of the image divides the input image in M squared blocks $S_m(x_s, y_s)$ with fixed size. Then, ROI A is obtained by applying the following algorithm for all M blocks:

$$\text{ROI } A(x,y)|_{x_s,y_s} = \begin{cases} 1, & \text{if var}(S(x_s, y_s) \leq t_1), \\ 0, & \text{otherwise,} \end{cases} \quad (5.1)$$

where t_1 is an empirically estimated threshold value and x_s, y_s are the local coordinates in the considered block.

ROI B is then obtained by applying flood-fill and open morphological operators to ROI A.

5.2.2.2 Quality Assessment Based on Computational Intelligence Classifiers (Method QA)

The proposed quality assessment method based on computational intelligence classifiers (Method QA) first extracts a set of features from the input frames and then performs the quality estimation using a neural classification system. First, a set of

45 features is computed for each frame, and then a feature selection technique is applied to increase the classification accuracy by selecting the most discriminative subset of features. A brief description of the initial feature vector F is reported. We refer to I_a as the region of image I pertaining to ROI A and to I_b as the region of image I pertaining to ROI B.

- $F(1)$: The area of ROI A.
- $F(2)$: The area of ROI B.
- $F(3)$: The ratio of $F(1)/F(2)$.
- $F(4)$: The standard deviation of I_a.
- $F(5)$: The standard deviation of I_b.
- $F(6)$: The standard deviation of the gradient phase of I_a.
- $F(7)$: The standard deviation of the gradient module of I_a.
- $F(8)$: The standard deviation of the gradient phase of I_b.
- $F(9)$: The standard deviation of the gradient module of I_b.
- $F(10)$–$F(11)$: The focus function $f_f(\cdot)$ approximated by a first-order polynomial. These features represent the focus level of the touchless fingerprint image, which is estimated from the local gradient of a set of reference points selected on edges and ridges pertaining to ROI B. The focus function $f_f(\cdot)$ is computed as follows:

  - The Sobel operator [182] is applied to I_b, thus obtaining the gradient module G_M and the gradient phase G_P.
  - The histogram H_{GM} of G_M is computed.
  - The cumulative frequency f_{cum} of H_{GM} is calculated using the following formulas:

  $$f(j) = \frac{H_{GM}(j)}{\left(\sum_{i=0}^{255} H_{GM}(i)\right)}, \tag{5.2}$$

  $$f_{cum} = \sum_{j=0}^{i} f(j). \tag{5.3}$$

  - The set of reference points $C_M(x, y)$ is estimated as follows:

  $$C_M(x, y) = G_M(x, y) \geq t, \tag{5.4}$$

  where $t = \arg\max_{0 \leq i \leq 255}(f_{cum}(i))$.
  - A subset of points R_{CM} is randomly selected from the set C_M.
  - For each element of R_{CM}, a local region is defined as a segment $s(i)$ centered in (x, y), with a fixed length angle normal to the corresponding gradient phase $G_P(x, y)$.
  - The histogram H_s of the gradient module G_M in the coordinates of every segment $s(i)$ is computed.

- Finally, the focus function $f_f(\cdot)$ is computed as follows:

$$f_f(\cdot) = \sum_{j=0}^{255} f(H(j)).\qquad(5.5)$$

- ∎ $F(12)$–$F(14)$: The coefficients of the second-order polynomial that approximates $f_f(\cdot)$. The focus function $f_f(\cdot)$ is computed in the same manner as features $F(10)$ and $F(11)$.
- ∎ $F(15)$–$F(24)$: The Fourier features obtained by analyzing the discrete Fourier transform of I_a. The computation of these features can be divided into the following steps:
  - All rows of I_a are concatenated to create a linear vector V_C.
  - The discrete Fourier transform of V_C is calculated, thus obtaining V_F.
  - The zero-frequency component of V_F is shifted into the central position of the vector, thus obtaining a vector V_S of N_F elements.
  - Finally, each Fourier feature is computed as follows:

$$\text{Fourier feature} = \sum_{j=N_F-N}^{N_F-1} |V_S(j+1) - V_S(j)|,\qquad(5.6)$$

  where N is the number of evaluated frequencies.
  The values of N used are $N = 50, 100, 150, 200, 250, 300, 350, 400, 450, 500$.
- ∎ $F(25)$–$F(34)$: The Fourier features of I_b, computed by evaluating the frequencies $N = 50, 100, 150, 200, 250, 300, 350, 400, 450, 500$. These features are computed in the same manner as the features from $F(15)$ to $F(24)$.
- ∎ $F(35)$: This feature represents the normalized gray-level differences in I and is computed as follows:

$$\Delta = \frac{\max(I) - \min(I)}{\max(I) + \min(I)}.\qquad(5.7)$$

- ∎ $F(36)$: The modulation of I_b.
- ∎ $F(37)$: The signal-to-noise ratio (SNR) of I_a, which is calculated as μ/σ, where μ and σ are the mean and standard deviation of I_a, respectively.
- ∎ $F(38)$: The SNR of I_b.
- ∎ $F(39)$: The Gabor feature of I_a. This feature aims to detect the presence of ridges in the touchless image, thus providing a measure of the focus level. The considered Gabor feature is the mean of the standard deviation of the eight matrices obtained by applying eight Gabor filters with different orientations (with angles equal to $[0, \pi/8, 1/4\pi, 3/8\pi, 1/2\pi, 5/8\pi, 3/2\pi, 7/8\pi]$) to I_a.

- $F(40)$: The Gabor feature of $I_b(x, y)$, which is computed as for the feature $F(39)$.
- $F(41)$: The mean of the local entropy of I_a. The image L_E representing the local entropy is created by first dividing I_a in local squared regions of 9×9 pixels, and then computing the local entropy of every region. Starting from the histogram H of a local region centered in (x, y), the local entropy is computed as follows:

$$L_E(x, y) = -\sum_{i=0}^{255} H_{(x,y)} \times \log_2(H_{(x,y)}(i)).$$ (5.8)

- $F(42)$: The mean of the local entropy L_E of I_b.
- $F(43)$: The standard deviation of the local entropy L_E of I_a.
- $F(44)$: The standard deviation of the local entropy L_E of I_b.
- $F(45)$: The global entropy of I_a.

The feature extraction step aims to present different parameters for estimating quality descriptors of touchless images. Depending on the application scenario, different subsets of the presented features will yield the best results in terms of accuracy and speed.

Then, feature reduction techniques are used to search the most discriminative feature subset with respect to the considered samples. One evaluated method is based on wrapper algorithms [315,316]. Other analyzed techniques are based on greedy-feature selection algorithms (such as Sequential Forward Selection, Sequential Backward Selection) and a custom wrapper [317].

Finally, neural networks are used to estimate the frame quality. Neural classifiers are adopted because they enable accurate learning of the nonlinear relationship between the computed feature set and the final frame quality. In this context, artificial neural networks are compared with other well-known classification techniques, such as k-nearest neighbor (kNN) and linear/quadratic discriminant classifiers.

5.2.2.3 Quality Assessment Based on Techniques Described in the Literature (Method QB)

This method first computes an image E representing the ridge pattern extracted from the touchless fingerprint image I. Then, the image quality is estimated by applying a well-known method in the literature designed for touch-based samples.

To permit real-time performance, the image E is computed by applying a fast enhancement algorithm to I_b. The enhancement consists of applying a filter with mask M in the frequency domain. This task is performed as follows:

$$E = \mathcal{F}^{-1}(\mathcal{F}(I_b(x, y) \cdot M)).$$ (5.9)

The mask $M(u, v)$ is obtained as follows:

$$M(u, v) = \exp\left\{\frac{(u^2 + v^2)}{r}\right\} - \exp\left\{\frac{(u^2 + v^2)}{\alpha_q r_q}\right\},$$ (5.10)

where u and v are the x and y spatial frequencies, respectively, in the frequency domain, r_q is a parameter empirically tuned according to the mean ridge frequency of the used dataset, and α_q is the spatial frequency of the frequency filter.

Then, the final estimation of the frame quality is given by analyzing the enhanced image E using the software NIST NFIQ (described in Section 4.5), as follows:

$$Q_{\text{NFIQ}} = \text{NIST NFIQ}(E(x, y)). \qquad (5.11)$$

The filters described in Equation 5.9 and Equation 5.10 obtain images representing only the spatial wavelengths of the ridge pattern, which are compatible with the approaches designed to estimate the quality of touch-based fingerprint samples. We selected the software NIST NFIQ because it is a standard reference method in the literature. The classifier returns five integer values. The value 1 is assigned to the best image quality.

5.2.3 Computation of Touch-Equivalent Images

The approach proposed for computing touch-equivalent fingerprint images from touchless acquisitions can be divided into two steps: image enhancement and resolution normalization.

Image enhancement is applied to obtain gray-scale images representing only the ridge pattern and to remove the noise present in touchless fingerprint images. We propose two image enhancement techniques designed for different application contexts. These techniques only consider gray-scale images; thus these techniques can be used for a variety of acquisition setups. Method EA is designed for particularly noisy images and permits many non-idealities due to uncontrolled acquisition conditions to be overcome. Method EB is designed to be more computationally efficient than Method EA. Moreover, Method EB can be applied to systems that can capture good-quality touchless images, obtaining results that are more accurate than those of Method EA.

The last step is the resolution normalization, which produces touch-equivalent images compatible with biometric algorithms designed for touch-based fingerprint images by normalizing the enhanced image to a resolution of approximately 500 ppi. This step assumes that the fingerprint images are captured at a constant distance from the camera.

5.2.3.1 Enhancement Based on Contextual Filters (Method EA)

The proposed enhancement technique based on contextual filters (Method EA) is designed to enhance fingerprint images captured in uncontrolled application contexts and can be divided into four distinct steps:

1. ROI estimation
2. Enhancement of the ridge visibility
3. Enhancement of the ridge pattern
4. Image binarization

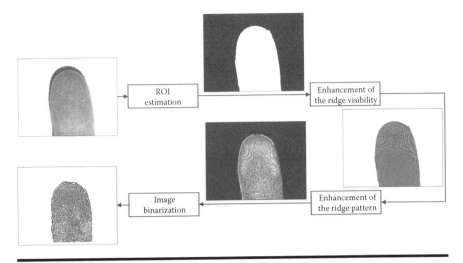

Figure 5.4 Schema of the method for enhancing touchless fingerprint images based on contextual filters (Method EA).

Figure 5.4 presents the schema of the proposed method.

The ROI is first estimated by applying the algorithm described in [168], which computes a binary image by evaluating the local variance of the image. Shadows are then removed by a threshold operation.

The second step is ridge-visibility enhancement. First, the skin texture is estimated and considered part of the background. The background image B_I is computed by applying a morphological opening operation with a mask s to the image I. Then, the image I_R representing the ridges is obtained as $I_R = I - B_I$. To reduce the noise present in the image, nonlinear equalization is performed as $I_L(x, y) = \log(I_R(x, y))$.

The third step is ridge pattern enhancement, which is based on a contextual filtering technique similar to the one proposed in [168], which applies a set of Gabor filters to the fingerprint image. The parameters of the filters are tuned according to the local ridge frequency and local ridge orientation. The result is the image I_E.

Finally, the image I_E is binarized. This task reduces the noise present in the ridge pattern representation, thus reducing the probability of estimating false minutiae in the sequent steps of the biometric recognition process. To improve the image contrast, the logarithm of I_E is computed as $I_I(x, y) = \log(I_E(x, y))$. Then, the binary image I_B representing the enhanced ridge pattern is computed as follows:

$$I_B(x, y) = \begin{cases} 0 & \text{if } I_I(x, y) \leq \operatorname*{argmax}_i(H(i)), \\ 1 & \text{otherwise}, \end{cases} \tag{5.12}$$

where H is the histogram of I_I.

5.2.3.2 Enhancement Based on the Ridge-Following Approach (Method EB)

The proposed enhancement technique based on the ridge-following approach (Method EB) is computationally more efficient and accurate than Method EA. However, this method can only be used in applications that can capture good-quality touchless images. The enhancement method can be divided into the following three steps:

1. ROI estimation
2. Ridge-visibility enhancement
3. Ridge-pattern enhancement and binarization

Steps 1 and 2 are performed using the same algorithms adopted by Method EA.

Then, noise reduction is performed by applying an eighth-order Butterworth low-pass filter with frequency f_l and size $d_l \times d_l$.

In the third step, ridge-pattern enhancement and binarization are performed using the software NIST MINDTCT [186], which is based on a ridge-following algorithm that directly computes the binary image I_B by analyzing the shape of every ridge of I_L.

5.2.3.3 Resolution Normalization

The proposed resolution-normalization technique assumes that the touchless fingerprint images are captured at a constant distance from the camera.

First, this method estimates the resolution of the captured images by evaluating the size of the plain captured at a distance Δ_L from the camera and then normalizes the touchless image to a resolution of 500 ppi.

Considering the plane perpendicular to the camera and at a distance Δ_L from the CCD, r_x inches along the horizontal direction of this plane correspond to i_x pixels along the horizontal direction of the captured images. Then, the normalization factor is estimated as follows:

$$n_f = \frac{i_x}{(r_x * 500)}. \tag{5.13}$$

5.2.4 Analysis of Level 1 Features in Touchless Fingerprint Images

A Level 1 analysis technique is used to estimate the position of the core point in touchless fingerprint images. The proposed method is described in [318].

The accuracy of the core localization is a critical factor that strongly affects the overall accuracy of many biometric systems that require a common reference point to perform fingerprint recognition. In these systems, all subsequent processing steps are related to the position of the core point.

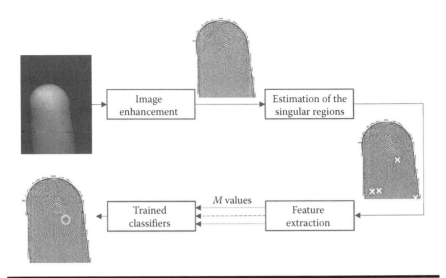

Figure 5.5 Schema of the proposed core-selection method. (R. Donida Labati et al., Measurement of the principal singular point in contact and contactless fingerprint images by using computational intelligence techniques, in *Proceedings of the IEEE International Conference on Computational Intelligence for Measurement Systems and Applications*, pp. 18–23. © 2010, IEEE.)

The proposed approach selects the core from a list of singular regions obtained by the Poincaré method [172]. This process is particularly difficult for touchless fingerprint samples acquired under uncontrolled conditions because the Poincaré method may estimate several false singularities. When the fingerprint image does not have singular regions, the maximum curvature point of the image is selected. The proposed approach uses computational intelligence techniques to estimate the position of the core point in touchless fingerprint images. This method can also be applied to touch-based images with satisfactory results.

The schema of the proposed approach is presented in Figure 5.5 and can be divided into the following four steps:

1. Image enhancement
2. Singular region estimation
3. Feature extraction
4. Core-point estimation

First, image enhancement reduces the noise present in the touchless image and improves the contrast between ridges and valleys. This step is performed using Method EA (Section 5.2.3).

Then, a list of the singular regions is obtained using an algorithm based on the Poincaré technique. In the case of fingerprints that do not present singularities, this algorithm returns the coordinates of the maximum curvature regions.

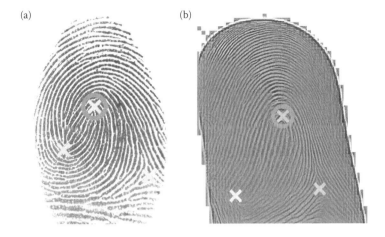

Figure 5.6 Application of the proposed approach for identifying the core point in fingerprint samples of the following: (a) a touch-based fingerprint image and (b) a touchless fingerprint image. The X markers represent the singular regions, and the O markers represent the core points. (R. Donida Labati et al., Measurement of the principal singular point in contact and contactless fingerprint images by using computational intelligence techniques, in *Proceedings of the IEEE International Conference on Computational Intelligence for Measurement Systems and Applications*, pp. 18–23. © 2010, IEEE.)

The feature extraction is then performed. An important feature is the fingerprint class, which is estimated using the software NIST PCASYS [186]. The considered classes are: arch, left loop, right loop, scar, tented arch, and whorl. This feature is particularly important for the selection of the core point because it provides detailed information on the global curvature of the ridge pattern by using a single class index. Other extracted characteristics include the coordinates of the singular regions and features describing the amount of information present in the local regions centered in these points.

The last step is the estimation of the coordinates of the core point using computational intelligence methods to select the best point from a set of singular regions by analyzing the computed features. Figure 5.6 presents the application of this approach to fingerprint samples acquired using a touchless system and a touch-based sensor.

5.2.4.1 Singular Region Estimation

An algorithm based on the Poincaré technique is used to compute a list of the coordinates of the singular regions and the type of every singularity. The proposed algorithm can be divided into the following steps:

- The ridge orientation map Ω is created using the technique presented in [172].
- The coordinates of the point $c_{ROI}(x, y)$ are calculated as the centroid of the ROI.
- The Poincaré technique is applied to I, thus obtaining the map of the Poincaré indices P.
- A binary image B_L describing the possible positions of loop points is computed as follows:

$$B_L(x, y) = \begin{cases} 0 & \text{if } 180 - t_L < I_P(x, y) < 180 + t_L, \\ 1 & \text{otherwise,} \end{cases} \qquad (5.14)$$

where t_L is an empirical threshold value.
- A binary image B_D describing the possible positions of delta points is computed as follows:

$$B_D(x, y) = \begin{cases} 0 & \text{if } -180 - t_D < I_P(x, y) < -180 + t_D, \\ 1 & \text{otherwise,} \end{cases} \qquad (5.15)$$

where t_D is an empirical threshold value.
- The centroid of each 8-connected region B_L is computed and stored in the vector of V_L.
- The centroid of each 8-connected region B_D is computed and stored in the vector of V_D.
- The vectors V_L and V_D are sorted by the distance from $c_{ROI}(x, y)$.
- The vector of the singular regions S is created by concatenating V_L and V_D.
- If no singular regions have been detected, then the maximum value of P nearest to $c_{ROI}(x, y)$ is considered a singular region, and its coordinates are input into S.

5.2.4.2 Feature Extraction

Two different features are computed from the input images. The first feature consists of the fingerprint classification obtained using the software NIST PCASYS [221]. This software is based on neural networks and returns an integer between 1 and 6 that describes the fingerprint class: (1) arch, (2) left loop, (3) right loop, (4) scar, (5) tented arch, and (6) whorl. The software NIST PCASYS also returns the probability of the fingerprint classification, which is also used as a feature of the proposed approach.

For each image, the vector V_{SP} contains all the singular regions of the fingerprint sample. First, V_{SP} is sorted by the type of singularity (loop and delta), and then the vector is sorted again by evaluating the Euclidean distance between the coordinates of the singular region and the centroid of the ROI $c_{ROI}(x, y)$.

The subsequent classification step requires feature sets with fixed length. Therefore, only the first M singular regions are selected and used in the subsequent computational steps. If the algorithm estimates the presence of additional singular regions far from the centroid of the ROI, these regions can be considered false positives with high probability and are thus omitted from further processing.

For each of the selected singular regions, the following features are computed:

- The coordinates of the singular region (x_O, y_O) in the coordinate space of the original image.
- The coordinates of the singular region (x_P, y_P) expressed as a percentage of the size of the ROI.
- The position of the singular region expressed in polar coordinates (ρ_O, θ_O) centered at $c_{ROI}(x, y)$.
- The angle θ_R of the singular region with respect to the fingerprint orientation α_f, which is computed as follows:

$$\theta_R = \theta_O - \alpha_f, \tag{5.16}$$

where α_f is the angle between the main bisector of the ROI and the y-axis.
- The mean and the standard deviation of the intensity in the circular region of radius r centered at the coordinates (x, y) of the singularity. These features describe the amount of information in the local area of the singular region.
- The coordinates of the centroid of the ROI $c_{ROI}(x, y)$.
- The angle of the orientation of the fingerprint α_f.
- The index of the singular region closest to the upper limit of the phalanx.
- The index of the singular region closest to $c_{ROI}(x, y)$.

For each image, the feature vector is composed of $7 + 11 \times M$ elements.

5.2.4.3 Core Estimation Using Computational Intelligence Techniques

The features computed for each fingerprint image are used as inputs for computational intelligence classifiers, which estimate the index of the singularity corresponding to the core point. Different classifiers have been considered. Additional details are discussed in the Section 6.1.2.

5.2.5 Analysis of Level 2 Features in Touchless Fingerprint Images

As described in Chapter 4, most touch-based fingerprint recognition systems in the literature are based on the analysis of Level 2 features. These methods usually estimate the minutiae coordinates and angles and then perform matching by searching the corresponding points in two or more templates. In most cases, this step is based on analyzing the Euclidean distance between pairs of points.

However, touchless biometric systems based on acquisition setups that do not use finger placement guides can capture images with different magnifications due to different placements of the finger during the acquisition step. Therefore, matching techniques based on evaluating the distances between minutiae points should be performed only after a resolution normalization step.

The proposed approach for evaluating Level 2 features of touchless fingerprint images is based on well-known algorithms. First, minutiae features are extracted from touch-equivalent images by applying the method described in Section 5.2.3. The minutiae estimation technique uses NIST MINDTCT software [186]. The minutiae on the border of the finger silhouette are removed because these minutiae are artifacts introduced by the image enhancement. Then, the resulting templates are compared using NIST BOZORTH3 software [186]. Details regarding the NIST MINDTCT and NIST BOZORTH3 algorithms are provided in Chapter 4.

5.2.6 Reduction of Perspective and Rotation Effects

We studied an approach for perspective deformation and roll rotation registration in touchless fingerprint recognition systems based on a single CCD camera [319]. This method is designed to work in uncontrolled applications, such as recognition systems integrated in mobile devices.

Perspective distortions in touchless fingerprint images can drastically decrease the overall accuracy of the biometric recognition system. One of the most important sources of distortion is the positioning of the finger during the acquisition process, which can present rotations with respect to the camera optical axis (Figure 5.7). This problem is particularly important in biometric systems that do not use finger placement guides.

Most of the matching algorithms in the literature produce poor results on images affected by perspective distortions because they are based on an evaluation of the Euclidean distances between minutiae points [5,6].

The proposed approach estimates the roll-angle difference between biometric samples by using neural networks and a set of features designed to detect differences in finger placements. A rotation compensation strategy is then applied by rotating a synthetic three-dimensional model of the finger shape.

During the enrollment phase, the proposed biometric recognition approach stores n_θ templates obtained from fingerprint images created by performing n_θ rotations of a synthetic three-dimensional model of the finger shape. These templates are computed using the minutiae-based method described in Section 5.2.5. Additional features are also stored for use during the verification phase to identify the best template for comparison with the fresh biometric data. These features consist of 18 real values obtained by analyzing both the finger shape and the ridge characteristics.

The same features are extracted from the fresh acquisition during the verification phase. A feature set is then created from the characteristics extracted from the fresh sample and the stored templates. The obtained data are used to estimate the angular distance between the compared touchless fingerprint acquisitions. Finally,

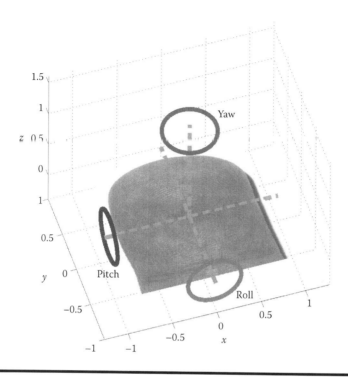

Figure 5.7 Possible rotations of the finger with respect to the camera optical axis in touchless recognition systems. (R. Donida Labati et al., Contactless fingerprint recognition: A neural approach for perspective and rotation effects reduction, in *Proceedings of the IEEE Workshop on Computational Intelligence in Biometrics and Identity Management,* **pp. 22–30. © 2013, IEEE.)**

the estimated angle is used to select the stored template to be matched with the fresh fingerprint acquisition.

The proposed biometric recognition approach is presented in Figure 5.8 and can be divided into the following steps:

1. Image preprocessing
2. Simulation of finger rotations
3. Feature extraction
4. Rotation estimation with neural networks
5. Template computation
6. Matching

5.2.6.1 Image Preprocessing

This step creates a touch-equivalent image I_B from a touchless image I. First, image enhancement is performed by applying Method EB (Section 5.2.3). Then, to obtain

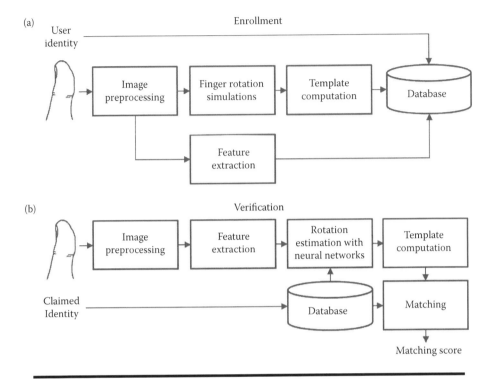

Figure 5.8 Biometric recognition process based on the studied approach for reducing perspective and rotation effects: (a) enrollment and (b) verification. This proposed approach can also be used in identification systems. (R. Donida Labati et al., Contactless fingerprint recognition: A neural approach for perspective and rotation effects reduction, in *Proceedings of the IEEE Workshop on Computational Intelligence in Biometrics and Identity Management*, pp. 22–30. © 2013, IEEE.)

images with a resolution of 500 ppi, resolution normalization is performed using the algorithm described in Section 5.2.3.

5.2.6.2 Finger Rotation Simulation

The proposed method for simulating finger rotations computes a synthetic three-dimensional model of the finger shape from the information extracted from the ROI image I_R. The ridge pattern is then superimposed on the obtained three-dimensional surface. Finally, the three-dimensional model of the fingerprint is rotated in three-dimensional space and a frontal sketch is created to obtain a synthetically rotated fingerprint image.

The three-dimensional model of the finger shape Z_s is a depth map created by analyzing the shape of the ROI. A third-order polynomial p is used to approximate the finger curvature along the z-axis. The height z of every row of Z_s is proportional to the width of the same row in I_R.

The polynomial p is imposed to pass through the points $(x_{min}, 0)$, $(x_m - x_m \cdot c_W, c_H)$, $(x_m + x_m \cdot c_W, c_H)$, $(x_{max}, 0)$, where x_{min} is the left limit of I_R along the horizontal axis, x_{max} is the right limit of I_R along the horizontal axis, and $x_m = (x_{max} + x_{min})/2$, c_W and c_H are values empirically estimated from the analysis of real three-dimensional fingerprint models.

To obtain a first estimation of the finger height, the polynomial p is fitted in the interval from x_{min} to x_{max}, obtaining the vector C.

Then, the height of every row i of Z_s is computed as follows:

$$Z_s(i) = I_R(i) \times C \times (X_{min}(i) - X_{max}(i)), \tag{5.17}$$

where X_{min} and X_{max} are vectors describing the x coordinates of the left and right limits of the ROI at every column i. Figure 5.9a shows an example of simulated finger shape.

The fingerprint model is completed by superimposing the touch-equivalent image I_B on the depth map Z_s.

A rigid transformation is then applied to Z_s, rotating the synthetic model by an angle τ, as follows:

$$Z_\tau = Z_s \begin{vmatrix} 1 & 0 & 0 \\ 0 & \cos(\tau) & -\sin(\tau) \\ 0 & \sin(\tau) & \cos(\tau) \end{vmatrix}, \tag{5.18}$$

where θ_R is the horizontal angle measured clockwise.

The synthetically rotated fingerprint image I_τ is computed as the frontal sketch of the rotated fingerprint model. The x and y coordinates of I_B are set as equal to the novel x and y coordinates of Z_τ. A bilinear interpolation is then applied to the image I_B to obtain a uniformly sampled matrix.

An example of a three-dimensional model of a rotated fingerprint is presented in Figure 5.9b, and the corresponding frontal sketch I_τ is presented in Figure 5.9c.

5.2.6.3 Feature Extraction

The estimation of the finger rotation is based on the analysis of a set of perspective features P_F, which are computed for each fingerprint image during the enrollment and verification phases. The feature set P_F comprises a measurement of the finger silhouette asymmetry δ and a matrix G_F of characteristics extracted by applying Gabor filters with different orientations to the touch-equivalent image I_B.

A compensation of the pan angle is performed before computing the silhouette asymmetry. The ROI image I_R is rotated along its centroid, minimizing

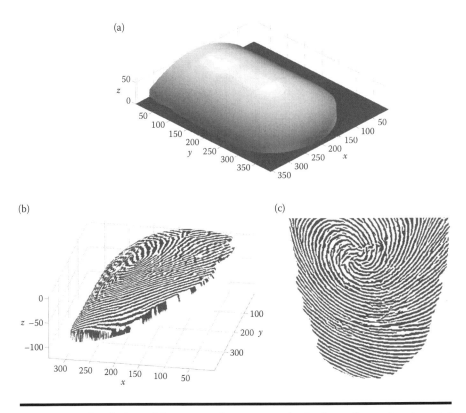

Figure 5.9 Simulation of finger rotations: (a) synthetic three-dimensional model of the finger shape; (b) rotated fingerprint model; and (c) resulting image. (R. Donida Labati et al., Contactless fingerprint recognition: A neural approach for perspective and rotation effects reduction, in *Proceedings of the IEEE Workshop on Computational Intelligence in Biometrics and Identity Management*, pp. 22–30. © 2013, IEEE.)

horizontal asymmetry:

$$\begin{cases} \Delta_A(\psi,y) = R_{\text{edge}}(\text{rot}(I_R,\psi),y) - L_{\text{edge}}(\text{rot}(I_R,\psi),y), \\ \hat{\psi} = \arg\min_\psi \left(\sum_{y=a_R}^{b_R} |\Delta(\psi,y)| \right), \end{cases} \tag{5.19}$$

where $\Delta(y) = R_{\text{edge}}(y) - L_{\text{edge}}(y)$; $L_{\text{edge}}(y)$ and $R_{\text{edge}}(y)$ are the left and right edges of the rotated ROI mask, respectively; $\text{rot}(\cdot)$ is the rotation function; a_R and b_R are the limits of the ROI along the y-axis; and $\hat{\psi}$ is the angle that minimizes the displacement of the finger silhouette.

Once the compensation of the pan angle has been performed, the remaining horizontal asymmetry in the first part of the ROI is evaluated as follows:

$$\delta = \sum_{y=a_R}^{c_R} |\Delta(\hat{\psi}, y)|, \tag{5.20}$$

where $c_R = \lfloor \kappa_R b_R \rfloor$, with $c_R > a_R$, which is an integer row index in the center of the rotated ROI images. Proper values of the parameter κ_R are those that permit the estimation of the horizontal asymmetry in the first third of the finger silhouette.

The analysis of the global characteristics of the ridge pattern (Level 1 analysis) can also be used to detect perspective distortions in touchless fingerprint samples. Therefore, Gabor filters (Equation 4.3) with different orientations ψ are applied to the image I_B to compute a set of 32 features G_P. Two images G_ψ are created by using filters with orientations $(0°, 90°)$. The set of features G_P is then obtained by computing the absolute average distance (AAD) of the intensity of every filtered image G_ψ divided into 4×4 rectangular regions. The feature matrix G_P is thus composed of $4 \times 4 \times 2$ AAD values.

5.2.6.4 Rotation Estimation with Neural Networks

This step uses neural networks to estimate a discrete value describing the roll-angle difference Δ_{theta} between the touchless fingerprint acquisitions A and B. First, the feature sets P_{FA} and P_{FB} are used to compute a feature vector F_V, as follows:

$$F_V = d_f(P_{FA}, P_{FB}), \tag{5.21}$$

represents the data fusion technique employed. The value of Δ_{theta} is finally estimated by the following:

$$\Delta_{theta} = NN(F_V), \tag{5.22}$$

where $NN(\cdot)$ is a feed-forward neural network.

- *Feature set computation*

  The feature set F_V is computed as an 18-element feature vector obtained from the feature sets P_{FA} and P_{FB} created from the acquisitions A and B, respectively. Every feature set P_F comprises the value δ and the $(4 \times 4 \times 2)$ matrix G_F.

  First, the Gabor features G_{FA} and G_{FB} are used to compute the 16-element matrix G_d describing the differences in Level 1 features in the acquisitions A and B. For each column i of the matrix G_{FA}, the corresponding values of $G_d(i)$ are obtained from the following:

  $$G_d(i, 1, 2) = interp(G_{FA}(i) - G_{FB}(i)), \tag{5.23}$$

  where $interp(\cdot)$ represents a linear approximation function.

Finally, the feature vector F_V is obtained as follows:

$$F_V = \delta_A, \delta_B, G_d(1), \ldots, G_d(16). \qquad (5.24)$$

■ *Rotation-difference estimation*

The difference in roll angle between touchless fingerprint acquisitions is difficult to estimate for single-camera systems. The fingers of different individuals vary greatly in size and shape. Moreover, touchless images can feature noise, shadows, and reflections.

In this context, the generalization capability of neural networks enables a robust estimation of the roll-angle difference between two touchless samples with very limited computational resource requirements compared to methods based on traditional algorithms.

The proposed approach is based on neural classifiers because it estimates the roll-angle difference as a value representing a discrete range of angles.

5.2.6.5 Template Computation

In the enrollment phase, the proposed method for the simulation of finger rotations is used to compute n_τ synthetically rotated images. Then, a set of n_τ minutiae templates $\bar{T}$ are computed and stored in the biometric database.

In the verification phase, the fresh acquisition A is used to create the minutiae template T_A, which is then compared only with the template $\bar{T}_B(\Delta_\tau)$. The discrete value of Δ_τ is the roll-angle difference estimated using the proposed neural-based method.

As described in Section 5.2.5, the matching score is computed using the software NIST BOZORTH3 [186].

5.3 Methods Based on Three-Dimensional Models

This section presents the studied approaches for the acquisition and elaboration of touchless three-dimensional fingerprint samples. First, a method for the three-dimensional reconstruction of minutiae points is presented. Then, three different methods for the three-dimensional reconstruction of the finger surface are discussed. Feature extraction and matching techniques based on the three-dimensional coordinates of minutiae points are then described. A method for the computation of touch-equivalent images by unwrapping fingerprint three-dimensional models is presented, and a technique for the quality evaluation of touch-equivalent images computed from three-dimensional models is finally described.

5.3.1 Three-Dimensional Reconstruction of Minutiae Points

A method for the three-dimensional reconstruction of minutiae points has been proposed [320]. This method permits the identification of corresponding minutiae

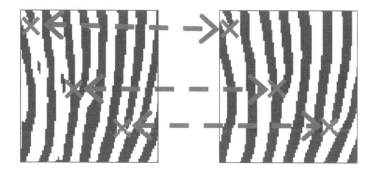

Figure 5.10 Example of three minutiae pairs in two different views of the same finger. (R. Donida Labati et al., A neural-based minutiae pair identification method for touchless fingerprint images, in *Proceedings of the IEEE Workshop on Computational Intelligence in Biometrics and Identity Management.* © 2011, IEEE.)

pairs in touchless fingerprint images and the estimation of their three-dimensional coordinates. In particular, this method is based on a novel minutiae matcher that uses neural networks and local features to search corresponding minutiae in touchless samples. Figure 5.10 shows an example of corresponding minutiae pairs detected by the proposed matcher; the arrows indicate correspondences between three minutiae pairs that are ready to be transformed into three-dimensional points.

The proposed minutiae matcher is compatible with images captured by multiple-view acquisition setups or with frame sequences capture by a single camera.

The first step of the proposed approach is image preprocessing, which segments and enhances the touchless fingerprint image. Second, the minutiae are extracted using a method designed for touch-based samples. Then, local features centered on each minutia are computed. The candidate minutiae pairs are then compared using trained neural classifiers, and the resulting pairs of points are used to estimate a template representing the three-dimensional coordinates of the minutiae. A matcher that compares three-dimensional features can then be used to compute the matching score within two templates. Figure 5.11 presents the schema of the proposed approach.

5.3.1.1 Acquisition, Image Preprocessing, and Minutiae Estimation

The schema of the acquisition setup employed is shown in Figure 5.12. The setup consists of two cameras placed at a distance Δ_H from the finger. The parameter α is the tilt angle of the cameras. The values of Δ_H and α are empirically tuned for each performed experiment. The illumination technique consists of a white LED.

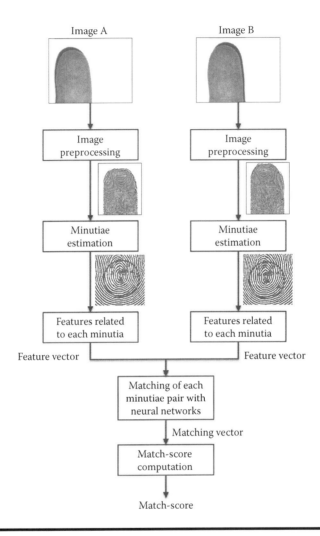

Figure 5.11 Schema of the proposed method for identifying corresponding minutiae points in touchless fingerprint images using computational intelligence techniques. (R. Donida Labati et al., A neural-based minutiae pair identification method for touchless fingeprint images, in *Proceedings of the IEEE Workshop on Computational Intelligence in Biometrics and Identity Management.* © 2011, IEEE.)

The preprocessing step consists of the segmentation and enhancement of the fingerprint images. First, the input images are segmented using the method described in Section 5.2.2. Then, the enhancement step is performed using Method EA (Section 5.2.3) to reduce the noise and increase the contrast of the ridge pattern.

Similar to the method described in Section 5.2.5, the minutiae are extracted using the software NIST MINDTCT [186].

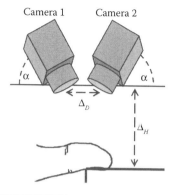

Figure 5.12 Schema of the proposed acquisition setup for computing three-dimensional minutiae points.

5.3.1.2 Computation of Features Related to Each Minutia

Three different matrices of local features (M_A, F_A, H_A) are extracted for a fingerprint image A. These features are computed by evaluating local regions centered in every minutia i.

The matrix M_A is composed by the following values obtained by the software NIST MINDTCT. For each minutiae point i, M_A comprises the following values:

- $M_A(i, 1)$: The x-coordinate.
- $M_A(i, 2)$: The y-coordinate.
- $M_A(i, 3)$: The angle α_m of the minutia normalized considering the median value of the ridge orientation map. This feature is computed as $\alpha_m = \beta + \theta$, where β is the median ridge orientation map and θ is the angle returned by the minutiae extractor.
- $M_A(i, 4)$: The quality index.

The matrix F_A is composed of N_F elements, which represent local FingerCode templates computed in local regions centered at every minutiae point. The templates are obtained by using an algorithm similar to that presented in [180], which can be divided into the following steps:

- Definition of a local circular ROI with a fixed radius r_G centered at the coordinates (x_i, y_i) of the minutiae point i.
- Tasseling of the local ROI in N_R rings and N_A arcs, thus obtaining $N_S = N_R \times N_A$ sectors S_i.
- Computation of a set of N_f Gabor filters with different orientations using 4.3, followed by application to the local area of the fingerprint image. This task creates N_f filtered images $T_{k\theta}$.
- Creation of the matrix $F_A(k, \theta)$ by computing the AAD 4.8 of each local region of the filtered images $T_{k\theta}$.

For each minutia, the FingerCode vector comprises $N_F = N_S \times N_f$ elements. Because the Fingercode template is not rotational-invariant, the feature matrix F_A is created according to the median ridge orientation β.

The matrix H_A is created by computing a set of histogram of oriented gradients (HOG) features in an area of fixed size ($N_p \times N_p$ pixels) centered at the coordinates of each minutia i. HOG features are descriptors of the image gradient features and are widely used in different pattern recognition applications. The algorithm used to compute HOG features is described in [321] and can be divided into the following steps:

- Computation of the gradient module G_M and the gradient phase G_P of the image I.
- Division of the images $G_M(x, y)$ and $G_P(x, y)$ into $c_w \times c_h$ cells.
- Quantization of the gradient orientation G_p of each cell into c_b orientation bins, weighted by the corresponding gradient magnitude G_M.
- Computation of a histogram with c_b orientations for each cell.

For each minutia i, the number of created HOG features is equal to $c_w \times c_h \times c_b$.

5.3.1.3 Classification of Minutiae Pairs

Computational intelligence techniques are used to estimate whether pairs of minutia extracted from images captured at different viewpoints represent the same point of the fingertip. For each minutiae pair, the proposed method thus performs a binary classification ("corresponding points," "noncorresponding points"). The method exploits the learning capability of feed-forward neural networks to perform the classification task. Additional details are reported in the experimental results section (Section 6.2).

5.3.1.4 Three-Dimensional Reconstruction of Minutiae Pairs

A metric reconstruction of the three-dimensional coordinates of the minutiae pairs is computed using a calibrated acquisition system [301]. Therefore, the z-coordinate of each three-dimensional point is estimated using the following triangulation formula:

$$z = \frac{fT}{x_A - x_B}, \tag{5.25}$$

where f is the focal length, T is the distance between the cameras, and x_A and x_B are corresponding points expressed in the two-dimensional coordinates of the images A and B, respectively.

The calibration of the multiple-view setup is performed off-line using the methods described in [322,323]. These methods use planar objects with distinctive reference points (the corners of a chessboard) to estimate the parameters describing the acquisition setup.

5.3.2 Three-Dimensional Reconstruction of the Finger Surface

The methods described in the literature for computing three-dimensional fingerprint samples require complex and constrained setups. These methods can be based on multiple cameras, special mirrors, and structured-light illumination systems. These methods all require the finger to be exactly placed and remain motionless during the time required to perform the biometric acquisition.

We present a novel approach for obtaining a three-dimensional reconstruction of the fingertip under less-constrained conditions than those conditions proposed in the literature. In fact, this approach permits three-dimensional fingerprint samples to be obtained without using finger placement guides and is based on the capture of multiple-view images in a single instant.

The following three methods for computing three-dimensional fingerprint samples based on different acquisition setups are proposed:

■ *Three-Dimensional Method A*: The three-dimensional reconstruction is based on a two-view acquisition system and a static projected pattern. The use of the projected pattern aims to reduce the time required to compute the three-dimensional models.

■ *Three-Dimensional Method B*: The acquisition setup comprises a two-view acquisition system and a white LED light. This setup is designed for integration in low-cost applications.

■ *Three-Dimensional Method C*: This method captures fingerprint images using a two-view acquisition technique and uses a diffuse blue light to enhance the visibility of the ridge pattern.

The technologies and algorithms employed in three-dimensional Method A and three-dimensional Method B are presented in [324].

The algorithms and acquisition setups differ among these proposed three-dimensional reconstruction methods; however, a common computational schema can be defined as follows:

1. Camera calibration
2. Image acquisition
3. Image preprocessing
4. Reference point extraction and matching
5. Cross-correlation matching of corresponding pairs of points
6. Reference point pair refinement
7. Three-dimensional surface estimation and image wrapping
8. Texture enhancement

The schema of the proposed three-dimensional reconstruction approach is presented in Figure 5.13.

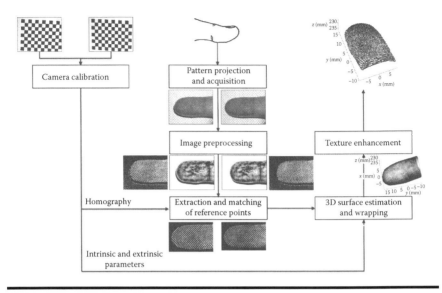

Figure 5.13 **Schema of the proposed method for the three-dimensional reconstruction of the finger volume. (R. Donida Labati et al., Fast 3-D fingertip reconstruction using a single two-view structured light acquisition, in *Proceedings of the IEEE Workshop on Biometric Measurements and Systems for Security and Medical Applications*, pp. 1–8. © 2011, IEEE.)**

5.3.2.1 Camera Calibration

The acquisition system is calibrated off-line before fingertip image capture using the methods described in [322,323]. These methods are based on planar objects with distinctive reference points (the corners of a chessboard) to estimate the intrinsic and extrinsic parameters of the cameras.

5.3.2.2 Image Acquisition

The three proposed acquisition setups are presented in Figure 5.14. Three-dimensional Method A is based on a fixed-projected pattern and enables fast three-dimensional reconstructions. Three-dimensional Method B only uses a white LED light to enhance the visibility of the ridge pattern for use in low-cost applications. Three-dimensional Method C uses a diffuse blue light to enhance the visibility of the ridge pattern.

5.3.2.2.1 Three-Dimensional Method A

Figure 5.15 presents a portion of the projected pattern. The colors of the chessboard are computed starting from a uniform RGB image:

$$C_R(x, y) = C_G(x, y) = C_B(x, y) \ \forall \ (x, y), \tag{5.26}$$

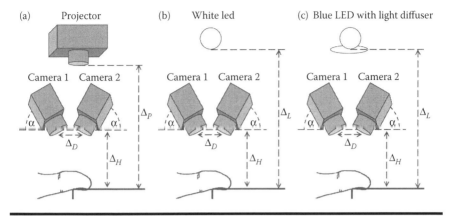

Figure 5.14 Schema of the proposed acquisition setups for computing three-dimensional fingerprint models: (a) three-dimensional Method A, (b) three-dimensional Method B, and (c) three-dimensional Method C.

Figure 5.15 Portion of the alternate-square projected pattern used by three-dimensional Method A. (R. Donida Labati et al., Fast 3-D fingertip reconstruction using a single two-view structured light acquisition, in *Proceedings of the IEEE Workshop on Biometric Measurements and Systems for Security and Medical Applications*, pp. 1–8. © 2011, IEEE.)

where C_R, C_G, and C_B are the channels of the image. The green and blue squares are obtained by summing and subtracting a constant value Δ_P from the C_G and C_B channels of the projected image:

$$\begin{cases} G_P(x,y) = C_G(x,y) + \Delta_P \\ B_P(x,y) = C_B(x,y) - \Delta_P \end{cases} \text{if } (x,y) \in \text{Green square,}$$

$$\begin{cases} G_P(x,y) = C_G(x,y) - \Delta_P \\ B_P(x,y) = C_B(x,y) + \Delta_P \end{cases} \text{if } (x,y) \in \text{Blue square,} \qquad (5.27)$$

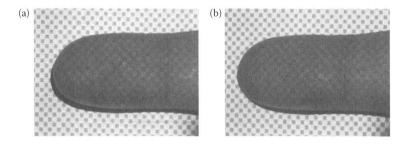

Figure 5.16 A two-view acquisition obtained using the proposed structured-light pattern (three-dimensional Method A). (R. Donida Labati et al., Fast 3-D fingertip reconstruction using a single two-view structured light acquisition, in *Proceedings of the IEEE Workshop on Biometric Measurements and Systems for Security and Medical Applications*, pp. 1–8. © 2011, IEEE.)

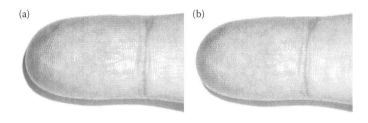

Figure 5.17 A two-view acquisition obtained using a white LED light (three-dimensional Method B).

where G_P and B_P are the green and blue channels of the resulting chessboard pattern, respectively (Figure 5.15). Figure 5.16 presents an example of finger images obtained from the two system cameras using the described projected pattern.

5.3.2.2.2 Three-Dimensional Method B

Three-dimensional Method B captures color images using a punctiform light source. The primary advantage of this acquisition setup is the low-cost illumination technique, which allows the acquisition systems to be used in different application contexts. However, the captured images are subject to problems associated with reflections and shadows. Figure 5.17 presents an example of captured pairs of images.

5.3.2.2.3 Three-Dimensional Method C

Three-dimensional Method C is based on a diffuse blue-light illumination. This type of light enhances the visibility of the ridge pattern, reducing problems related to reflections, and shadows compared with three-dimensional Method B. In fact,

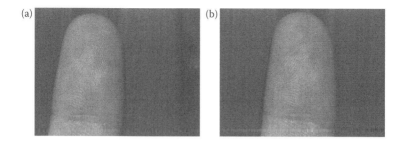

Figure 5.18 Two-view acquisition obtained using the proposed blue light illumination technique (three-dimensional Method C).

as described in Chapter 4, the touchless fingerprint images captured using a diffuse blue light are usually of better quality than the images captured using a white light source. An example pair of images is presented in Figure 5.18.

5.3.2.3 Image Preprocessing

The image preprocessing step is different for different proposed three-dimensional reconstruction methods. In fact, the images obtained using three-dimensional Method A contain additional information compared with the images captured by three-dimensional Method B and three-dimensional Method C.

5.3.2.3.1 Three-Dimensional Method A

The images obtained using three-dimensional Method A are more complex than the images captured by the other methods and require the ridge pattern F to be separated from the square pattern P.

The ridge pattern F is computed using the following equation:

$$I_F(x,y) = C_G(x,y) + C_B(x,y), \tag{5.28}$$

where C_G and C_B are the green and blue channels of the captured image, respectively. An example of the ridge pattern image F is presented in Figure 5.19a.

The first approximation of the square pattern I_P is computed as follows:

$$I_P(x,y) = C_G(x,y) - C_B(x,y). \tag{5.29}$$

To reduce the presence of noise, a histogram equalization is performed, and then a logarithm image is computed as follows:

$$P(x,y) = \log(1 - I_P(x,y)). \tag{5.30}$$

The resulting image representing the projected pattern P is shown in Figure 5.19b.

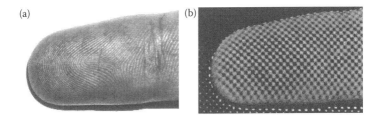

Figure 5.19 Separation of finger details and structured-light pattern in images captured using the proposed three-dimensional Method B: (a) finger ridge pattern *F* and (b) enhanced structured-light pattern *P*. (R. Donida Labati et al., *Fast 3-D fingertip reconstruction using a single two-view structured light acquisition*, in *Proceedings of the IEEE Workshop on Biometric Measurements and Systems for Security and Medical Applications*, pp. 1–8. © 2011, IEEE.)

Fingerprint segmentation is then performed to remove both the background and shadows. A binary image of the finger shape is estimated as follows:

$$I_s(x, y) = \begin{cases} 1 & t_l < F(x, y) < t_h, \\ 0 & \text{otherwise}, \end{cases} \tag{5.31}$$

where t_l and t_h are two threshold values. To remove areas of the segmented region that do not pertain to the finger, the obtained binary mask is then treated by a morphological opening operation, followed by a closing operation. The structural element used is a circle with radius r_s. Because the external areas of touchless finger images are frequently affected by noise, a morphological erosion is also performed by using a circular structural element with radius r_e.

5.3.2.3.2 Three-Dimensional Method B

The images captured by three-dimensional Method B only represent the finger skin. Therefore, the ridge pattern image *F* is obtained by converting the captured image in gray-scale. Figure 5.20a presents an example of the results obtained.

Image segmentation is then performed using the same algorithm adopted in three-dimensional Method A.

5.3.2.3.3 Three-Dimensional Method C

The images obtained by three-dimensional Method C represent the fingerprint illuminated by a blue light. In these images, the details of the ridge pattern are particularly visible in the blue channel of the RGB colorspace. Therefore, the ridge-pattern image *F* is considered channel *B* of the captured image. An example of the obtained results is presented in Figure 5.20b.

(a)   (b)

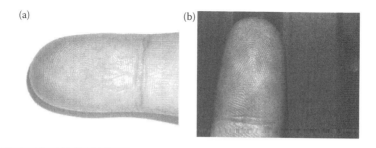

Figure 5.20 Ridge pattern image examples obtained using different acquisition and preprocessing techniques: (a) three-dimensional Method B and (b) three-dimensional Method C.

In contrast to the images obtained by three-dimensional Method A and three-dimensional Method B, the images captured by three-dimensional Method C have a dark background. Therefore, the segmentation step does not consider the presence of finger shadows. First, the binary image I_S is computed as follows:

$$I_s(x,y) = \begin{cases} 1 & F(x,y) > t_O, \\ 0 & \text{otherwise}, \end{cases} \quad (5.32)$$

where t_O is obtained by applying Otsu's method [182]. Then, the ROI is obtained by eroding the image I_S. The structural element used is a circle with radius r_e.

5.3.2.4 Extracting and Matching the Reference Points

The method described in Section 5.3.1 allows pairs of corresponding points to be obtained in multiple-view images of touchless fingerprints. However, this method can obtain a limited number of corresponding pairs of points because this method only considers the minutiae present in the fingerprint images. Moreover, fingerprints do not present a fixed number of minutiae, and the minutiae extraction algorithms can improperly estimate the coordinates of the feature points. Therefore, a three-dimensional reconstruction of the finger shape based only on the corresponding minutiae points can produce poor-quality samples. To overcome these problems, we propose two techniques to estimate a greater number of corresponding points in touchless fingerprint images.

The first proposed technique (F Correlation) can be applied in all of the proposed reconstruction methods because this technique only considers the fingerprint pattern F images. The second technique (P Correlation) can only be used by three-dimensional Method A because this technique uses the supplementary information related to P to accelerate the search for corresponding pairs of points.

The two proposed approaches are based on cross-correlation. A cross-correlation coefficient r is computed according to the following formula:

$$r = \frac{\sum_m \sum_n (A_{mn} - \bar{A})(B_{mn} - \bar{B})}{\sqrt{(\sum_m \sum_n (A_{mn} - \bar{A})^2)(\sum_m \sum_n (B_{mn} - \bar{B})^2)}}, \quad 1 < m < l, \ 1 < n < l, \qquad (5.33)$$

where A and B are squared local regions of size $l \times l$.

5.3.2.4.1 F Correlation

First, the image F'_A is computed by rectifying F_A such that corresponding epipolar lines of F_B (captured by Camera B) lie along horizontal scan lines [325]. This task limits both the computational complexity and the probability of false matches. Rectification is performed using the homography matrix H estimated during the calibration of the multiple-view acquisition setup.

Then, a set of reference points is selected in the image F'_A by resampling the ROI with a step of s_d pixels.

For each reference point (x'_A, y'_A), the matching point in the image F_B is identified by applying a block-matching technique that searches the maximum normalized cross-correlation between a $l \times l$ squared region centered in x'_A, and a $l \times l$ squared region sliding in the corresponding epipolar line of F_B. The search region consists of $\pm w$ pixels along the epipolar line (x-axis of the image F'_A). The point x_B with the maximum normalized correlation value r corresponds to x_A.

5.3.2.4.2 P Correlation

This method can only be applied to images obtained using three-dimensional Method A because this method uses the information related to the projected pattern. The centroids of the black-and-white squares of the images B_A and B_B are estimated to obtain a set of candidate matching points (Figure 5.21).

Similar to the matching algorithm F Correlation, this method computes the rectified image F'_A first and then identifies the pairs of corresponding points by computing the normalized maximum normalized correlation r. To reduce the computational time, the search for corresponding points is performed considering only the centroids of the projected squares. To limit possible errors, the described method is applied by first considering only the centroids of the white squares, followed by the centroids of the black squares.

5.3.2.5 Refining the Pairs of Reference Points

Pairs of erroneously matched points are identified before computing the three-dimensional model. The following strategies are adopted to remove the outliers:

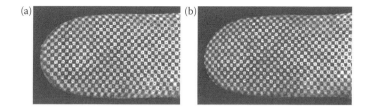

Figure 5.21 **Extracted reference points in three-dimensional Method A: (a) reference points in the first image (Camera 1) and (b) reference points in the second image (Camera 2). (R. Donida Labati et al., Fast 3-D fingertip reconstruction using a single two-view structured light acquisition, in *Proceedings of the IEEE Workshop on Biometric Measurements and Systems for Security and Medical Applications*, pp. 1–8. © 2011, IEEE.)**

5.3.2.5.1 Three-Dimensional Method A

Erroneously matched points are identified by estimating a homography between the matched pairs of points using the RANSAC algorithm [326]. Pairs of points that do not fit the obtained homography (according to an empirically estimated threshold) are considered false correspondences and thus discarded.

Then, a spike filter based on a statistical approach is applied. This algorithm is based on evaluating the Euclidean distance between corresponding points. The distance related to a pair of points is compared with the local mean and standard deviation of the distances computed for every pair of points as follows:

$$\bar{D}_S - \sigma_{DS} - t_{1s} < d(x_A, y_A, x_B, y_B)_i < \bar{D}_S + \sigma_{DS} + t_{s1} \quad \forall 1 < i < N, \quad (5.34)$$

where $\bar{D}_S$ and σ_{DS} are the weighted mean and the standard deviation of the Euclidean distance, respectively, between each pair of points (x_A, y_A) and (x_B, y_B) in the local region S_A. The weights used to compute the weighted mean are the inverse of the Euclidean distances between (x_A, y_A) and each point of the region S_A.

5.3.2.5.2 Three-Dimensional Method B

First, an outlier removal task is performed. The pairs of matched points of V_x related to the same column of F_A are stored in a vector V_C. This vector is interpolated using a third-order polynomial, obtaining the vector V_{CI}. For each element of the vector, $V_x(i)$ and $V_y(i)$ are removed if $|V_C(i) - V_{CI}(i)| < t_{s2}$, where t_{s2} is a fixed threshold.

The values of D_X and D_Y are obtained by the linear interpolation of V_x and V_y in each point pertaining to the ROI of the image A. The coordinates of each point (x_B, y_B) corresponding to (x_A, y_A) are obtained using the following equations:

$$\begin{aligned} x_B &= x_A + D_x(x_A, y_A), \\ y_B &= y_A + D_y(x_A, y_A). \end{aligned} \quad (5.35)$$

Then, the previously described spike filter (Equation 5.34) is adopted.

5.3.2.5.3 Three-Dimensional Method C

A thin-plate spline is applied to the set of corresponding points. This task allows a smooth and accurate representation of the finger surface to be obtained. However, the thin-plate spline interpolation can only obtain good results for data with low noise levels.

5.3.2.6 Three-Dimensional Surface Estimation and Image Wrapping

This step estimates the dense three-dimensional shape of the finger. Considering the limited resolution of the cameras used, this method does not estimate the roughness of the fingertip but computes a three-dimensional representation of the finger and wraps an image of the ridge pattern on the estimated model.

First, the depth coordinate of each three-dimensional point is estimated using the triangulation formula (Equation 5.28). Then, a dense three-dimensional representation of the finger is computed. Two equispaced maps M_x are M_y are computed and the depth map M_z, which describes the coordinates of the points of the finger, is then obtained by using a linear interpolation with a constant step s_{interp}. The texture of the three-dimensional model M_T is obtained by applying the same interpolation algorithm on the points of the image F_A. An example of a point cloud and its relative estimated surface are presented in Figure 5.22.

5.3.2.7 Texture Enhancement

Touchless fingerprint images captured by the proposed two-view acquisition setups are affected by the same non-idealities that are present in touchless images captured by single CCD cameras. However, the images captured by three-dimensional Method A and three-dimensional Method B contain more noise than images captured by three-dimensional Method C. Therefore, three-dimensional Method A and three-dimensional Method B enhance the texture MT using the algorithm EA (Section 5.2.3). Three-dimensional Method C uses the algorithm EB (Section 5.2.3).

5.3.3 Feature Extraction and Matching Based on Three-Dimensional Templates

A relevant study regarding three-dimensional minutiae matching is presented in [294]. The matching method represents every minutiae point as (x, y, z, θ, ϕ), where z is the minutiae coordinate in the z-axis and ϕ is the minutiae orientation in spherical coordinates with unit length 1. The matching score between two templates is computed by iteratively aligning the minutiae sets and by searching the number of corresponding minutiae pairs. Correspondences between minutiae

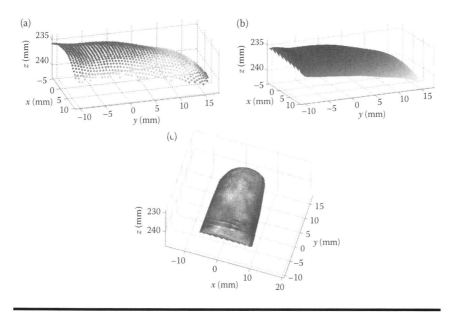

Figure 5.22 **Three-dimensional model obtained using the proposed method for computing the finger volume: (a) point cloud, (b) finger volume, and (c) three-dimensional model with wrapped texture. (R. Donida Labati et al., Fast 3-D fingertip reconstruction using a single two-view structured light acquisition, in** *Proceedings of the IEEE Workshop on Biometric Measurements and Systems for Security and Medical Applications,* **pp. 1–8. © 2011, IEEE.)**

points are determined by evaluating fixed thresholds. While the results of this study are promising, the accuracy of the three-dimensional matcher is lower than that of two-dimensional algorithms. Moreover, this method does not propose heuristics for efficiently aligning three-dimensional templates.

We have studied alternative techniques based on both Delaunay graphs and three-dimensional minutiae features. Delaunay graphs are used to reduce the number of iterations performed in the template alignment. An example of the template used is presented in Figure 5.23.

The performed tests produced encouraging results for different datasets. However, the evaluated method produced results that were less accurate than those obtained using well-known techniques based on two-dimensional features. The primary problem is related to the rigid registration of three-dimensional samples captured using different finger placements because well-known techniques designed for aligning three-dimensional models, such ICIP, can produce poor results when used with three-dimensional minutiae templates. In fact, these templates are composed of small sets of points (usually less than 100), can present false and missed minutiae, and are composed of points estimated with small errors. Thus, template-alignment strategies that are more robust must be designed.

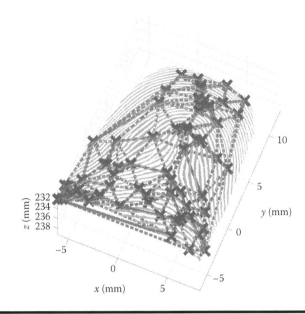

Figure 5.23 Graphical example of the template used by the proposed matching algorithm based on three-dimensional fingerprint models. The X symbols represent the minutiae, and the lines represent the triangles of the Delaunay graph.

5.3.4 Unwrapping Three-Dimensional Models

A method for computing touch-equivalent images from three-dimensional fingerprint samples has been studied [324]. This method is designed to obtain fingerprint images similar to rolled fingerprints and to be compatible with existing matching algorithms designed for touch-based fingerprint images. This method is parametric and approximates the finger shape using rings of different radii, similar to the algorithm described in [288].

Starting from the matrices M_x, M_y, M_z, which describe the three-dimensional shape of the fingerprint model, an approximate finger shape is first computed. For each y-coordinate, the coordinates of $M_x(y)$ and $M_z(y)$ are approximated to a circle by applying the Newton method, obtaining the vectors V_{x1} and V_{z1}, which represent the centers of the approximating circles.

Then, a spike filter is applied to remove potential outliers. The filter removes the elements of V_{x1} and V_{z1} with distances greater than the experimentally estimated thresholds t_x and t_z.

To obtain a smooth approximating shape, the values of V_{x1} and V_{z1} are then approximated using a first-order polynomial, obtaining V_x and V_z.

Next, the radii of the circles are estimated, thus obtaining the vector V_r. For each y coordinate of the three-dimensional model, the approximating radius $V_r(y)$ is obtained by computing the average distance between the center of the circle

$(V_x(y), V_z(y))$ and the actual coordinates of the three-dimensional model $M_x(i, y)$, and $M_z(i, y)$. The obtained radii are then smoothed by approximating the vector V_z to a third-order polynomial.

Finally, the approximated three-dimensional finger shape is mapped into two-dimensional space. For each y coordinate, the pixel intensities of the texture map M_T are mapped into the coordinate system of the corresponding approximating circle. The unwrapped fingerprint image is obtained by applying a linear interpolation with constant step to M_T in the novel Polar coordinates. To obtain images with a resolution of approximately 500 ppi along the y-axis, an interpolation step of 0.0508 mm is used. Similarly, a resolution of approximately 500 ppi along the y-axis is obtained by sampling every approximating circle with a constant step of 0.0508 mm along its perimeter.

Finally, the image I_U is segmented using an algorithm based on the local standard deviation of the image [168].

Figure 5.24 presents an example of the results of this method.

5.3.5 Quality Assessment of Touch-Equivalent Fingerprint Images Obtained from Three-Dimensional Models

An approach specifically designed for the quality evaluation of touch-equivalent fingerprint images obtained by unwrapping three-dimensional models has been developed and presented in [327].

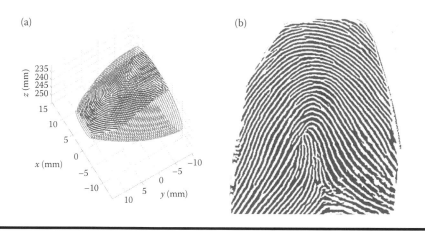

Figure 5.24 Example of results obtained by applying the proposed three-dimensional unwrapping technique: (a) three-dimensional fingerprint sample and (b) the resulting unwrapped image. (R. Donida Labati et al., Fast 3-D fingertip reconstruction using a single two-view structured light acquisition, in *Proceedings of the IEEE Workshop on Biometric Measurements and Systems for Security and Medical Applications*, pp. 1–8. © 2011, IEEE.)

Touch-equivalent images obtained by unwrapping three-dimensional finger-print models present different non-idealities compared with samples captured by touch-based sensors. This type of images can present low visibility of the ridges in local regions of the fingerprint due to external factors in the acquisition procedure and artifacts introduced by the three-dimensional reconstruction and unwrapping steps of the biometric recognition process.

Touch-equivalent images computed from three-dimensional models can present the following non-idealities:

1. Deformations of the ridge pattern caused by poor-quality regions of the three-dimensional finger shape
2. Artifacts caused by outlier points in the three-dimensional finger shape
3. Local regions with artifacts introduced by the image enhancement algorithm, which correspond to low-contrast regions of the captured images regions of the input images

Figure 5.25 presents examples of these non-idealities.

The presence of deformation and artifacts in fingerprint images can drastically decrease the accuracy of biometric systems. Quality assessment techniques should therefore increase the final accuracy and reliability of the biometric recognition process.

No methods in the literature have been specifically designed for evaluating touch-equivalent images obtained by unwrapping three-dimensional models.

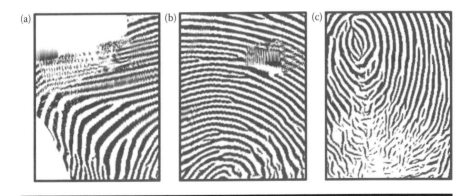

Figure 5.25 Examples of problems that can be present in unwrapped finger-print images: (a) a deformation caused by a badly reconstructed portion of the three-dimensional model, (b) an artifact caused by the presence of a spike in the three-dimensional model, and (c) an area with low visibility of the ridge pattern. (R. Donida Labati et al., Quality measurement of unwrapped three-dimensional fingerprints: A neural networks approach, in *Proceedings of the International Joint Conference on Neural Networks*, pp. 1123–1130. © 2012, IEEE.)

Therefore, a new approach for the quality evaluation of touch-equivalent fingerprint images obtained by unwrapping three-dimensional models is proposed. This approach uses computational intelligence techniques to evaluate the presence of non-idealities in the ridge pattern. The quality is expressed using discrete classes to easily discard low-quality touch-equivalent images during the biometric recognition process.

The proposed approach can be divided into the following steps:

1. Image segmentation
2. Feature extraction
3. Quality estimation using computational intelligence techniques

5.3.5.1 Image Segmentation

The ROI is computed as a binary image I_R, which is obtained by thresholding the standard deviation of the intensity in local areas of $l_r \times l_r$ pixels. The threshold t_s is empirically estimated based on the dataset employed.

5.3.5.2 Feature Extraction

The following sets of distinctive features are extracted:

- Minutiae features (F_M): These features consist of the number of minutiae and the mean and standard deviation of their quality. The minutiae are extracted using the software NIST BOZORTH3 [186].
- Features describing the ROI (F_S): The computed features consist of the length and width of the ROI and the eccentricity of its shape. This set of features is computed to detect errors in the computation of the three-dimensional finger shape.
- Gabor features (F_G): This set consists of a vector of $m_G \times n_G \times \theta_G$ values. First, θ_G Gabor filters 4.3 with different orientations are applied to touch-equivalent image I_E, obtaining θ_G images G_θ. Then, the AAD 4.8 is computed for each local rectangular region (with size $m_G \times n_G$ pixels) of every image G_θ.
- HOG features (F_H): This set consists of a vector of $c_w \times c_h \times c_b$ values, which is obtained by computing a set of HOG features in the ROI of the touch-equivalent image I_E. Similar to the method described in Section 5.2.2, we compute these features by applying the algorithm described in [321].
- Standard deviation of Gabor features ($F_{G\sigma}$): This set consists of a vector of $m_G \times n_G$ values. The feature extraction algorithm used to compute the set of Gabor features F_G is applied, obtaining $m_G \times n_G \times \theta_G$ values representing the local AAD of filtered images. For each local region, the computed feature is the standard deviation of the filter responses described by the corresponding

θ_G AAD values. These features are designed to evaluate the presence of information in the local regions of the image to be evaluated, thereby reducing the number of Gabor feature values used.

■ Standard deviation of HOG features ($F_{H\sigma}$): This set consists of a vector of $c_w \times c_h$ values. HOG features are computed by dividing the touch-equivalent image I_E into $c_w \times c_h$ local regions. For each local region, the computed feature is the standard deviation of the c_h obtained HOG values. The size of this feature set is smaller than that of the set F_H.

5.3.5.3 Quality Estimation

Our goal is to distinguish touch-equivalent images of sufficient quality for use in biometric recognition from poor-quality images affected by non-idealities due to noise introduced by the acquisition process and by the three-dimensional reconstruction and unwrapping algorithms. Quality estimation is thus considered a classification problem that categorizes touch-equivalent fingerprint images into "sufficient" and "poor." We use feed-forward neural networks to perform this classification to learn the complex nonlinear relationships between the input features.

5.4 Computation of Synthetic Touchless Fingerprint Samples

Studies regarding computing synthetic samples representing the three-dimensional ridge pattern are presented in [261], and techniques for creating three-dimensional models representing the full finger are described in [328].

The design of touchless fingerprint recognition systems requires the creation of biometric databases that are large enough to confirm the validity of the envisioned algorithms and acquisition techniques. This process requires substantial resources in terms of time and money. Moreover, this process can involve the realization of new acquisition techniques. To limit the effort necessary to collect biometric data, performing tests on simulated data can be useful. Some studies have examined the creation of synthetic fingerprint images that realistically resemble those images acquired using touch-based sensors. However, the only work in the literature regarding simulating three-dimensional fingerprint models aims to create three-dimensional phantoms using a three-dimensional printer [298]. Therefore, this work does not create images or three-dimensional data that simulate acquisitions under real operating conditions that can be used to evaluate acquisition setups and recognition algorithms. For example, this work does not consider the skin texture, illumination, or intrinsic and extrinsic parameters of acquisition setups based on multiple cameras. Thus, the proposed method could be a useful resource for researchers studying biometrics and three-dimensional reconstruction by facilitating the simulation of different environmental conditions and acquisition setups. Moreover, this method enables the creation of datasets of synthetic samples that can be used as a reference for comparing the performance of future algorithms.

The proposed method creates synthetic three-dimensional models of fingerprints acquired using different setups and light conditions using image processing algorithms and well-known techniques designed for feature extraction from touch-based fingerprint images. This method first infers information regarding the ridge pattern from a touch-based fingerprint image (live, rolled, or synthetic). Then, it models the three-dimensional finger shape. The three-dimensional shape of the ridge pattern is modeled and superimposed on the finger shape. Realism is increased by simulating pores, incipient ridges, camera focus, and noise. Then, the skin color and illumination conditions are introduced. Finally, images representing acquisitions performed by multiple-view camera systems are computed. This method can be divided into the following steps:

- Computation of the silhouette
- Computation of the three-dimensional finger shape
- Computation of the three-dimensional ridge pattern
- Simulation of the skin color
- Simulation of the illumination conditions
- Simulation of a multiple-view acquisition

5.4.1 Computation of the Silhouette

First, the studied method for computing synthetic three-dimensional samples computes the three-dimensional finger shape by simulating the finger silhouette starting from average models obtained from real touchless images. The use of knowledge extracted from real samples allows more realistic results to be obtained.

5.4.1.1 Computation of Average Silhouettes from Real Images

We consider three classes of fingers in the computation of the silhouette models: the thumb, little finger, and other fingers. The average silhouette is estimated offline once for each finger class.

For each finger class, the first step of the average finger silhouette computation is the segmentation of real touchless samples, which is performed using the technique described in Section 5.2.2.

Then, the segmented images pertaining to the same class are aligned. First, the top of the finger shapes are aligned along the y-axis. Then, the mean finger shape is estimated by analyzing the left and right halves of the fingerprint samples. For each row y, the mean points $m_L(y)$ and $m_R(y)$ of the finger halves F_L and F_R are

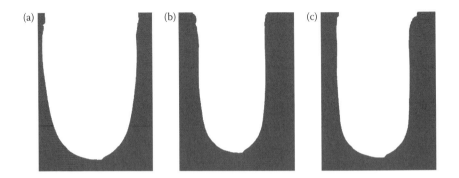

Figure 5.26 Average finger silhouettes: (a) average thumb silhouette, (b) average little finger silhouette, and (c) average silhouette of the remaining fingers. (R. Donida Labati et al., Accurate 3D fingerprint virtual environment for biometric technology evaluations and experiment design, in *Proceedings of the IEEE International Conference on Computational Intelligence and Virtual Environments for Measurement Systems and Applications*, pp. 43–48. © 2013, IEEE.)

estimated as follows:

$$m_L(y) = \frac{1}{n_C} \sum_{i=1}^{n_C} F_L(x_i, y),$$

$$m_R(y) = \frac{1}{n_C} \sum_{i=1}^{n_C} F_R(x_i, y) \forall y, \tag{5.36}$$

where n_C is the number of considered touchless fingerprint samples. Finally, the average silhouette is obtained by filling the area enclosed between the points $m_L(y)$ and $m_R(y)$. Examples of average finger silhouettes are presented in Figure 5.26.

5.4.1.2 Simulation of the Finger Silhouette

In the computation of every synthetic three-dimensional model, the finger class is first defined. Then, a touch-based fingerprint image I_T is selected to create a realistic three-dimensional synthetic sample based on its ridge pattern.

The finger silhouette S' is computed by extracting information from both the touch-based image I_T and the average silhouette $\bar{S}$ corresponding to the chosen finger class. The average silhouette $\bar{S}$ is resized based on the ratio between the maximum width of I_T and the maximum width of $\bar{S}$, obtaining the image S'. Finally, S' is cropped starting from the top of the last phalanx to fit the area of I_T.

5.4.2 Computation of the Three-Dimensional Finger Shape

This step estimates the three-dimensional shape of the finger from the silhouette S'. This method computes the finger shape using a parametric model of the finger in a manner similar to the technique described in Section 5.2.6. This parametric model has been obtained by studying three-dimensional models reconstructed from real touchless acquisitions.

The finger curvature is defined as the fourth order polynomial p_A passing by the following x and y coordinates:

$$\{(x_{min}, 0), (x_m - a_W x_m, a_{H1}), (x_m, a_{H2}),$$
$$(x_m + a_W x_{mean}, a_{H1}), (x_{max}, 0)\}, \tag{5.37}$$

where x_{min} is the left limit of S' along the x-axis, x_{max} is the right limit of S' along the x-axis, $x_{mean} = (x_{max} + x_{min})/2$, and a_W, a_{H1} and a_{H2} are parameters empirically estimated by analyzing three-dimensional finger models reconstructed from real acquisitions.

Similar to the algorithm described in Section 5.2.6, the finger curvature is described by the vector C, which is obtained by fitting the polynomial p_A in the interval (x_{min}, x_{max}). Every column of the depth map S_z is calculated as follows:

$$S_z(i) = S'(i) \times C \times (X_{min}(i) - X_{max}(i)), \tag{5.38}$$

where X_{min} and X_{max} are the vectors that represent the left and right limits of the finger silhouette S', respectively. Figure 5.27 presents an example of the computed finger shape.

5.4.3 Computation of the Three-Dimensional Ridge Pattern

This step computes the height of the ridges by analyzing the ridge pattern of a real or synthetic touch-based image I_T.

5.4.3.1 Ridge-Pattern Analysis

First, the ridge pattern is analyzed to estimate a map G that characterizes the heights of the ridges in nondimensional space. The proposed method considers the height of the ridges and valleys in proportion to their intensity in the image obtained by applying a contextual filtering technique based on Gabor filters 4.3. This technique is used to reduce the noise and obtain a smooth representation of the ridge shapes.

Then, a binary image B_r is computed to search the regions of the image G representing the ridges as follows:

$$B_r(x, y) = \begin{cases} 0 & \text{if } G(x, y) < t_b, \\ 1 & \text{otherwise,} \end{cases} \tag{5.39}$$

where t_b is an empirically estimated threshold.

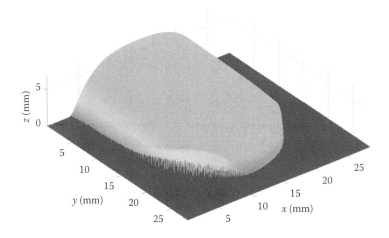

Figure 5.27 Example of a synthetic three-dimensional shape of the finger. (R. Donida Labati et al., Accurate 3D fingerprint virtual environment for biometric technology evaluations and experiment design, in *Proceedings of the IEEE International Conference on Computational Intelligence and Virtual Environments for Measurement Systems and Applications*, pp. 43–48. © 2013, IEEE.)

The next task is the computation of the matrices R and V, which represent an approximation of the z values of ridges and valleys, respectively. These matrices are obtained as follows:

$$R(x,y) = \log\left(G(x,y) \times (B_r(x,y))\right), \tag{5.40}$$

$$V(x,y) = \log\left(1 - G(x,y) \times B_r(x,y)\right). \tag{5.41}$$

The values of R are then mapped in a range from $(1 - \Delta_r)$ to 1, and the values of the valleys V are mapped in a range from 0 to Δ_r. The parameter Δ_r is empirically estimated by analyzing three-dimensional finger models reconstructed from real acquisitions.

Then, the matrices R and V are used to create the image H_r, which describes the height of the complete ridge pattern in non-dimensional coordinates, as follows:

$$H_r(x,y) = R(x,y) + V(x,y). \tag{5.42}$$

Finally, the contrast between the ridges and valleys is increased by applying an adaptive histogram equalization to H_r.

5.4.3.2 Noise Injection

To increase the realism of the simulated three-dimensional fingerprint, this step introduces pores, incipient ridges, and acquisition noise.

First, a binary image B_n of randomly distributed white pixels is created. Then, the image B_n is dilated using a mask of $n_n \times n_n$ pixels. Finally, the pores and incipient ridges are simulated by applying the following equation:

$$
\begin{aligned}
H_n(x,y) = H(x,y) &- \Delta_n(B_n(x,y)(B(x,y))), \\
&+ \Delta_n(B_n(x,y)B(x,y)),
\end{aligned}
\tag{5.43}
$$

where Δ_n is a constant value estimated from the analysis of real touchless acquisitions.

5.4.3.3 Computation of the Three-Dimensional Ridges

First, the height image H_n is smoothed as follows:

$$
H_q(x,y) =
\begin{cases}
t_w & \text{if } H_n(x,y) > t_w, \\
H_n(x,y) & \text{if } H_n(x,y) \le t_w,
\end{cases}
\tag{5.44}
$$

where t_w corresponds to the p_n percentile of the histogram of H_n. The value of p_n is empirically estimated.

5.4.3.4 Simulation of Camera Focus

To simulate the out-of-focus areas of the peripheral image regions, the fingertip pattern is progressively smoothed toward the borders.

First, five focus regions R_i, $i = 1, \ldots, 5$, are defined as follows:

$$
R_i = (S' \wedge (S' \ominus K_m(i))),
\tag{5.45}
$$

where $\ominus$ represents the morphological erosion operator and K_m is a set of erosion masks. The set of masks K_m employed increases in size: 1/15, 1/12, 1/9, 1/8, or 1/7 of S'.

For each focus region, a low-pass Gaussian filter is applied to H_q as follows:

$$
H_f(x,y) = H_q(x,y) * G_i \text{ if } (x,y) \in S_i \quad \forall i = 1, \ldots, 5,
\tag{5.46}
$$

where G_i is a Gaussian filter with size of $m_i \times m_i$ pixels and standard deviation σ_i. The parameters m_i and σ_i are empirically tuned by analyzing real touchless images, such as $m_i > m_{i+1}$ and $\sigma_i > \sigma_{i+1}$.

Figure 5.28 presents an example of a simulated camera focus on the ridge pattern.

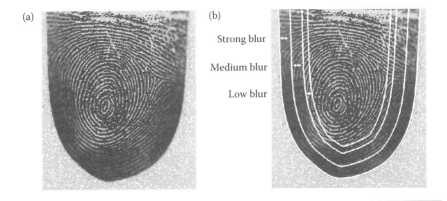

Figure 5.28 Graphical example of camera-focus simulation. For simplicity, only three regions are depicted: (a) a ridge pattern and (b) the ridge pattern in (a) blurred according to the various regions. (R. Donida Labati et al., Accurate 3D fingerprint virtual environment for biometric technology evaluations and experiment design, in *Proceedings of the IEEE International Conference on Computational Intelligence and Virtual Environments for Measurement Systems and Applications*, pp. 43–48. © 2013, IEEE.)

5.4.3.5 Superimposition of the Ridges

This step computes a depth map representing the final three-dimensional model by merging the finger shape and the three-dimensional ridge pattern.

First, the two-dimensional coordinates (x, y) of the finger shape S_z and the depth map of the ridge pattern H_q are converted to millimeters. This task is performed by assuming that S_z and H_q have the same resolution as the touch-based image I_T, equal to 500 DPI.

Then, the values of H_q are adjusted to a mean height of 0.06 mm, which corresponds to the mean height of human ridges.

The ridges on the fingertip surface are superimposed by summing the ridge height H_q to S_z along the direction of the surface normals. First, the normal vectors to every point $S(x, y)$ are computed, producing a three-element vector (N_x, N_y, N_z). The three-dimensional ridges are then superimposed as follows:

$$
\begin{aligned}
x_r &= x_s + N_x H_q(x, y), \\
y_r &= y_s + N_y H_q(x, y), \\
z_r &= z_s + N_z H_q(x, y)
\end{aligned}
\tag{5.47}
$$

Finally, a dense depth map $S_r(x, y)$ is computed by resampling the coordinates x_r and y_r with a constant step and by applying a bilinear interpolation to z_r to obtain the three-dimensional height in the new coordinate system.

5.4.4 Simulation of the Skin Color

The finger color is estimated from a real touchless fingerprint image I_l. First, a Gaussian low-pass filter is applied to remove the details of the ridge pattern from I_l, thus obtaining an image I_f representing only the underlying skin color and the vein pattern. This task is performed as follows:

$$I_f = I_l * G_l, \tag{5.48}$$

where G_l is a Gaussian filter with an area of $m_l \times m_l$ pixels and a standard deviation equal to the empirically estimated value σ_p.

Then, speckle noise is added using the following equation:

$$I_p = I_f + n_s \times I_f, \tag{5.49}$$

where n_s represents uniformly distributed random noise with mean equal to 0 and variance equal to v_s. Different values of the parameter v_s are empirically estimated to simulate different acquisition setups.

Subsequently, a texture-mapping algorithm is used to superimpose the noise image I_p onto the depth map S_r.

5.4.5 Simulation of the Illumination Conditions

This step simulates the light conditions by considering both the environmental illumination and an artificial light source. Moreover, the proposed illumination model realistically simulates the skin reflections.

The environmental light is simulated as a light source placed at infinity. The considered artificial light source is an LED illuminator, which is simulated as a punctiform light placed at coordinates (x_l, y_l, z_l) and radiating uniformly in all directions. The reflection of the finger surface is approximated using the Phong lighting algorithm. However, more complex rendering methods can be used [329].

5.4.6 Simulation of a Multiple-View Acquisition

The proposed method permits the simulation of images acquired by different configurations of multiple-view acquisition setups.

Starting from the intrinsic and extrinsic parameters of real calibration data obtained from the real acquisition setups described in Section 5.3.1.1, the proposed method computes three-dimensional projections as follows:

$$\begin{bmatrix} u & v & 1 \end{bmatrix}^{\mathrm{T}} = KR|t \begin{bmatrix} X & Y & Z & 1 \end{bmatrix}^{\mathrm{T}}, \tag{5.50}$$

where (u, v) are the pixel coordinates projected in the image coordinates, K is the matrix of the intrinsic camera parameters, R is the rotation matrix, and t is the translation vector.

Finally, the images related to every view are obtained by resampling the coordinates (u, v) with a constant step equal to 1, followed by an interpolation of the intensity values in the new coordinate system using a bilinear interpolation algorithm.

5.5 Summary

Different methods have been presented for all steps of touchless fingerprint recognition systems based on two-dimensional and three-dimensional samples. All methods are based on acquisition setups that can be considered less constrained than most of the systems described in the literature because these methods do not use finger placement guides and can capture samples in a single instant of time.

The studied techniques based on two-dimensional samples use acquisition setups based on single CCD cameras that capture fingerprint images at a distance of more than 200 mm from the sensor. Moreover, these techniques do not require complex and expensive illumination techniques.

To evaluate the presence of noise and other non-idealities in the touchless images, a quality assessment technique specifically designed for touchless fingerprint images has been studied. This technique is based on neural classifiers and discards samples of insufficient quality.

A technique for evaluating Level 1 characteristics in touchless images has been implemented. This technique permits an estimation of the coordinates of the core point using neural networks.

Level 2 analysis techniques have also been studied. In particular, a method for computing touch-equivalent images has been designed. This method can be used to perform biometric recognitions based on techniques designed for evaluating minutiae features in touch-based fingerprint images.

A technique analyzing perspective and rotation effects has also been implemented. This technique uses neural networks to estimate the acquisition angle and to compute synthetic models of the three-dimensional finger shape to compensate for perspective effects.

Biometric systems based on three-dimensional samples require acquisition setups and recognition algorithms that are more complex. Techniques for estimating the three-dimensional coordinates of minutiae points and the three-dimensional reconstruction of the finger surface have been studied. These techniques are based on multiple-view hardware setups and perform biometric acquisitions in a less-constrained manner than most systems described in the literature.

Feature extraction and matching techniques based on three-dimensional templates have also been presented.

Moreover, an unwrapping method for computing touch-equivalent images from three-dimensional samples and a technique for assessing the quality of the resulting fingerprint images have been studied.

A tool for simulating synthetic touchless fingerprint samples has also been developed to reduce the efforts necessary for acquiring the biometric samples required to test the proposed approaches.

Chapter 6

Experimental Results

This chapter presents the results of the evaluation of approaches for touchless fingerprint recognition based on two-dimensional and three-dimensional samples (Chapter 5).

First, a performance analysis of the proposed techniques based on two-dimensional samples is described. The presented quality assessment method for touchless fingerprint images (Section 5.2.2) is evaluated first, and then the results of the proposed core-point estimation technique for touchless fingerprint images are presented (Section 5.2.4). The accuracy of the studied approach for reducing perspective distortions and the effects of finger rotation is also presented (Section 5.2.6).

A quantitative evaluation of the accuracy of the presented three-dimensional reconstruction techniques (Sections 5.3.1 and 5.3.2) is also described. First, the accuracy of the method designed to reconstruct the three-dimensional coordinates of minutiae points is discussed. Then, the results obtained using different techniques for the three-dimensional reconstruction of the finger surface are compared.

Next, discussions of the results obtained by applying the proposed techniques for computing touch-equivalent images (Section 5.3.4) and the presented quality-assessment approach (Section 5.3.5) are presented.

Then, the proposed biometric recognition techniques are compared with touch-based recognition systems in a scenario evaluation. This comparison is based on the most common practices and figures of merit in the literature and includes different aspects of biometric technologies: (I) accuracy, (II) speed, (III) cost, (IV) scalability, (V) interoperability, (VI) usability, (VII) social acceptance, (VIII) security, and (IX) privacy.

Finally, the proposed technique for computing synthetic touchless fingerprint samples (Section 5.4) is analyzed.

This chapter is organized as follows: Section 6.1 presents the experiments performed using the proposed methods designed for elaborating touchless two-dimensional fingerprint samples. Section 6.2 compares different three-dimensional reconstruction techniques and presents results for the unwrapping of three-dimensional fingerprint models. A comparison of the proposed touchless recognition techniques with touch-based fingerprint recognition systems is presented in Section 6.3. Finally, a discussion of the accuracy obtained using the proposed approach for computing synthetic touchless samples is presented in Section 6.4.

6.1 Methods Based on Single Touchless Images

This section presents performance evaluations of the proposed methods for acquiring and elaborating two-dimensional fingerprint samples, described in Section 5.2. In particular, the results obtained using the presented quality-assessment method for touchless fingerprint images (Section 5.2.2), the approach for reducing perspective and rotation effects (Section 5.2.6), and the core-detection technique (Section 5.2.4) are provided.

6.1.1 Quality Estimation of Touchless Fingerprint Images

The proposed quality-assessment approach designed for touchless fingerprint images (Section 5.2.2) aims to identify the best-quality frames in frame sequences describing a finger moving toward the camera. We tested this approach using different biometric datasets collected in our laboratory and compared the results of the proposed Method QA and Method QB.

6.1.1.1 Creation of the Training and Test Datasets

No available public datasets include touchless fingerprint images captured under unconstrained conditions. Therefore, we created four different datasets to evaluate the performance of the proposed methods in different operating scenarios. Dataset Aq comprises 79 gray-scale frame sequences depicting different fingers captured using a Sony XCD-V90 camera. Every frame sequence has a duration of 6 s, and each frame has a size of 1920×1024 pixels. The frame rate was 30 fps. The illumination was controlled by a white LED, and the focal length was 25 mm. For each frame sequence, a user brought a finger near the camera, representing a movement of approximately 20 mm. The quality of every frame of the acquired sequences was labeled by an experienced supervisor in the following five categories:

■ $Q = 5$ (Poor): The ridge pattern is not detectable, or the frame does not include the ROI.

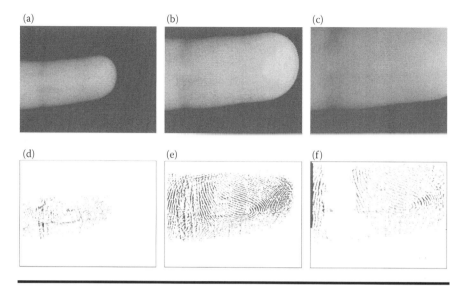

Figure 6.1 Method QB: Frame prefiltering. Subplots (a–c) present examples of frames with different quality levels, whereas subplots (d–f) present the output of the filter in the ROI region. The proposed filtering algorithm tends to enhance the focus of only the ridge portion and produces random-like patterns in the blurred regions. This behavior helps the subsequent NFIQ algorithm to properly estimate the quality of the frame. (R. Donida Labati, V. Piuri, and F. Scotti, Neural-based quality measurement of fingerprint images in contactless biometric systems, in *Proceedings of the International Joint Conference on Neural Networks*, pp. 1–8. ⓒ 2010, IEEE.)

■ Q = 4 (Fair): The contrast of the ridge pattern is poor because the frame is blurred.
■ Q = 3 (Good): The ridge pattern is visible, but the frame presents blurred regions.
■ Q = 2 (Very good): The ridge pattern is clearly visible.
■ Q = 1 (Excellent): The fingerprint image presents high contrast between the ridges and valleys. Figure 6.1 presents examples of classified frames.

Labeling the frames of Dataset Aq was a complex task because the defocusing effect is not proportional to the finger distance and because the frame sequences describe fingers moving with uncontrolled direction and speed.

We created three other datasets by subsampling Dataset Aq: Dataset Bq comprised five acquisitions from the same individual (993 frames); Dataset Cq comprised five frame acquisitions from five different individuals (997 frames); and Dataset Dq comprised 360 randomly selected frames captured under all evaluated operating conditions. For each dataset, we prepared a version classified into five quality categories and a version classified into two categories (Sufficient: quality class = 1; Insufficient: quality class > 1). The two-class datasets were created to produce a classifier capable of directly detecting the best frames to be used by the other modules of the biometric recognition system. In the following section, we refer to these datasets as Bq-5, Bq-2, Cq-5, Cq-2, Dq-5, and Dq-2.

6.1.1.2 Application of Proposed Method QB

The accuracy of Method QB is strictly dependent on the capability of the image enhancement algorithms (Section 5.9 and Section 5.10) to increase the contrast between ridges and valleys and minimize the noise and the effects of the other image components. The frequency filter used to enhance the touchless fingerprint samples was tuned on images captured using the setup presented in Figure 5.2, setting α_q equal to 10 and r_q equal to the height of the image. The output of this algorithm is a gray-scale image in which the in-focus regions of the input frame present enhanced visibility of the ridge pattern and blurred areas of the input frame are substituted by random-like patterns. This output facilitates the detection of out-of-focus regions, thus helping the subsequent algorithms to perform a robust estimation of the frame quality.

As a preliminary analysis, we evaluated the results obtained for estimating the quality of touchless fingerprint images using the five-classes approach proposed by the NIST. Figure 6.2 presents three examples of frames pertaining to Dataset Aq. Subplots (a)–(c) display a poor-quality frame, a good-quality frame, and an out-of-focus frame, respectively. Subplot (d) presents the output of Method QB (dots) compared with the quality labels assigned by the supervisor (dashed line) and feature $F(7)$ of Method QA.

The performed experiments indicate that Method QB produces poor results in estimating the quality of the best frames with respect to the labels set by the supervisor. Notably, proposed feature $F(7)$ properly follows the quality class assigned by the supervisor, displaying a smooth pattern. This feature exhibits the same behavior in all evaluated frame sequences.

6.1.1.3 Application of Proposed Method QA

The parameters of Method QA were empirically tuned on the available frame sequences to the following values: the size S was equal to 20×20 pixels, the threshold t_1 was equal to 0.05, and the size of the mask used to compute ROI B was equal to 30×30 pixels.

The produced primary Dataset Aq contains 14,220 frames; thus, managing this quantity of feature vectors with the best wrapper algorithms available in the literature is nearly impossible. Therefore, we used three subsampled datasets that describe different operating conditions of the biometric recognition system. The experiments indicated that the best performing feature selection technique for the considered datasets is forward selection [330]. The feature-selection step estimated the best feature combinations for the datasets Bq-5, Bq-2, Cq-5, Cq-2, Dq-5, and Dq-2.

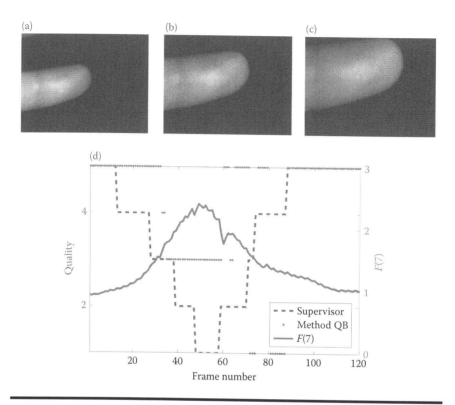

Figure 6.2 **Examples of the frame quality of a frame sequence describing a touch-less fingerprint acquisition: (a) poor-quality frame; (b) good-quality frame; (c) excellent quality frame; (d) graph representing the quality levels related to a frame sequence. This graph shows the labels selected by the supervisor (dashed line), the pattern of the feature F(7) of Method QA (continuous line), and the qual-ity levels produced by Method QB (dotted line). (R. Donida Labati, V. Piuri, and F. Scotti, Neural-based quality measurement of fingerprint images in contactless biometric systems, in** *Proceedings of the International Joint Conference on Neural Networks,* **pp. 1–8. © 2010, IEEE.)**

To examine the complexity of the considered classification problem, we exam-ined the following classifier families:

■ The linear Bayes normal classifier (LDC)
■ The quadratic Bayes normal classifier (QDC)
■ The k-nearest-neighbor classifier with different values of the parameter k (1, 3, and 5)
■ Feed-forward neural networks with a single hidden layer comprising different numbers of nodes

To adequately evaluate the generalization capability of the trained neural networks, the classifiers were validated by adopting the N-fold cross-validation technique with $N = 10$ [331]. The neural networks comprised a linear node for the output layer and a single hidden layer of log-sigmoidal nodes. We evaluated configurations of the neural networks with different number of nodes in the hidden layer. The neural classifiers were trained using the back-propagation algorithm.

6.1.1.4 Final Results and Discussion

The best subsets of features estimated by the forward-selection method are reported in Table 6.1. Table 6.2 compares the classification accuracy obtained by the

Table 6.1 **Feature Subsets Used to Evaluate the Implemented Quality Assessment Method**

Dataset	Feature Subset
Bq-5	[1, 2, 6, 7, 35, 40, 41, 45]
Bq-2	[1, 3, 6, 7, 8, 31, 37, 38, 42]
Cq-5	[1, 6, 7, 41, 42, 43, 45]
Cq-2	[2, 3, 6, 7, 29, 31, 36, 45]
Dq-5	[6, 8, 35, 45]
Dq-2	[4, 6, 10, 11, 12, 13, 16, 27, 33, 44]

Source: R. Donida Labati, V. Piuri, and F. Scotti, Neural-based quality measurement of fingerprint images in contactless biometric systems, in *Proceedings of the International Joint Conference on Neural Networks*, pp. 1–8. © 2010, IEEE.

Note: The datasets Bq-5, Cq-5, and Dq-5 are classified into five classes. The datasets Bq-2, Cq-2, and Dq-2 are classified into two classes.

Table 6.2 **The Accuracy of the Quality Assessment Method**

	Method QA								Method QB
	Linear		FFNN-3		FFNN-5		kNN		
Dataset	Mean	Std	Mean	Std	Mean	Std	Mean	Std	Mean
Bq-5	0.191	0.002	0.046	0.004	0.065	0.004	0.042	0.004	0.478
Bq-2	0.083	0.000	0.013	0.000	0.017	0.004	0.011	0.001	0.140
Cq-5	0.239	0.004	0.066	0.008	0.065	0.001	0.068	0.000	0.358
Cq-2	0.049	0.001	0.013	0.002	0.016	0.001	0.015	0.003	0.150
Dq-5	0.354	0.006	0.275	0.008	0.286	0.012	0.278	0.016	0.469
Dq-2	0.047	0.000	0.064	0.024	0.047	0.000	0.050	0.000	0.180

Source: R. Donida Labati, V. Piuri, and F. Scotti, Neural-based quality measurement of fingerprint images in contactless biometric systems, in *Proceedings of the International Joint Conference on Neural Networks*, pp. 1–8. © 2010, IEEE.

Note: Classification methods of Method QA: linear classifier (Linear), feed-forward neural networks with one hidden layer composed of three nodes (FFNN-3), feed-forward neural networks with a hidden layer composed of five nodes (FFNN-5), and kNN with $k = 1$ (kNN). The datasets Bq-5, Cq-5, and Dq-5 are classified into five classes. The datasets Bq-2, Cq-2, and Dq-2 are classified into two classes.

neural networks with those of the classical inductive classification methods for all considered datasets. These results pertain to the classifier configurations that achieved the best performances during the test (the number of nodes in the hidden layers for the neural networks, the value of k for the kNN classifiers, and the feature sets returned by the sequential forward-selection method).

Table 6.2 indicates that Method QA achieved remarkable accuracy for the considered datasets compared with Method QB. In addition, the neural-based classifier displayed higher accuracy than other classical inductive classification systems. Only the kNN classifiers displayed comparable accuracy (for the considered datasets); however, the neural classifiers were characterized by a significantly lower computational complexity. Table 6.3 presents the computational gain of the neural-based approach with respect to the kNN classification family. The minimum gain factor observed for all evaluated datasets was 46. The experiments demonstrated that the configuration of Method QA based on neural networks is the most appropriate for use in real-time application contexts.

The total computational time of the feature extraction step was also evaluated. Most of the features were computed in less than 0.05 s. The features $F(39)$ and $F(40)$ required approximately 2.3 s, and the features $F(41)$, $F(42)$, $F(43)$ and $F(44)$ required approximately 4 s. Features $F(10)$ and $F(11)$ were simultaneously computed in approximately 1.2 s. Similarly, computing features $F(12)$, $F(13)$, and $F(14)$ required approximately 1.2 s. The feature-selection step indicates that

Table 6.3 Computational Gain of the Implemented Quality Assessment Method

Dataset	Computational Gain
Bq-5	201.459
Bq-2	205.527
Cq-5	121.756
Cq-2	539.839
Dq-5	159.852
Dq-2	46.456

Source: R. Donida Labati, V. Piuri, and F. Scotti, Neural-based quality measurement of fingerprint images in contactless biometric systems, in *Proceedings of the International Joint Conference on Neural Networks*, pp. 1–8. © 2010, IEEE.

Note: The computational gain is based on the ratio of the computational time required by the most accurate traditional classifier to the computational time of the most accurate neural network. The accuracy values are reported in Table 6.2.

the most computationally expensive features can be replaced by simpler features without affecting the final accuracy of the quality assessment method, thus saving computational time.

The most computationally expensive step of Method QA is the ROI estimation. The algorithms for estimating ROI A and ROI B are not optimized and can require up to 1.5 and 3.7 s, respectively. In contrast, optimized implementations can reduce the computational complexity by a factor of 100. This point is crucial to achieving the real-time goal.

All results presented are based on scripts written in Matlab language (Version 7.6) and executed on an Intel Centrino 2.0 GHz in Windows XP.

Thus, the results obtained demonstrate that the presented method is feasible and that this method offers satisfactory accuracy in real-time applications with images captured at a distance of greater than 0.2 m.

6.1.2 Analysis of Level 1 Features in Touchless Fingerprint Images

In this section, we describe the tests performed to evaluate the proposed core-detection method designed for touchless fingerprint images (Section 5.2.4). We present the creation of the experimental datasets and the results obtained by the computational intelligence classifiers used by the proposed method for extracting the core point in touchless fingerprint images. Then, the obtained results are compared with those results obtained using heuristic methods.

6.1.2.1 Creation of the Training and Test Datasets

Two different datasets of fingerprint images were used for the training and testing steps: a set of touch-based images (Dataset Ac) and a set of touchless images (Dataset Bc). Figure 6.3 presents two examples of touch-based samples from Dataset Ac and two examples of touchless samples from Dataset Bc.

Dataset Ac comprised 498 samples acquired using a CrossMatch V300 sensor [332,333]. Eight images were acquired for each individual. The image resolution was 500 ppi, and the image size was 512×480 pixels.

Dataset Bc comprised 71 samples of different fingers captured using a Sony XCD-V90 camera. The acquisition setup employed is described in Section 5.2. The illumination was controlled by a white LED, the focal length was 25 mm, and the distance from the finger to the camera was approximately 200 mm. The acquired samples were gray-scale images with a dimension of 1920×1024 pixels.

We manually estimated the core position of every image for each dataset by selecting its position from the list extracted by the algorithm described in Section 5.2.4.

(a)   (b)   (c)   (d)

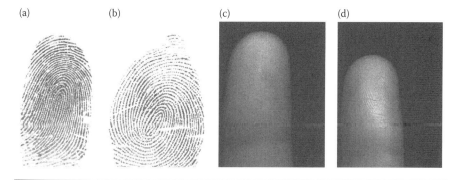

Figure 6.3 **Examples of images produced by touch-based (a, b) and touchless (c, d) sensors used in our experiments for the implemented method for estimating the core point. (R. Donida Labati, V. Piuri, and F. Scotti, Measurement of the principal singular point in fingerprint images: A neural approach, in *Proceedings of the IEEE International Conference on Computational Intelligence for Measurement Systems and Applications*, pp. 18–23. © 2010, IEEE.)**

6.1.2.2 Computational Intelligence Techniques and Obtained Results

Different classification paradigms were tested to further elucidate the complexity of the learning problem embedded in the datasets. In particular, we evaluated: the Fisher linear-discriminant classifier; the quadratic classifiers with different preprocessing methods; the k-nearest-neighbor classifier with odd values of the parameter k (1, 3, and 5); and feed-forward neural networks with different configurations of the hidden layer.

To identify the best configuration of the neural networks, we evaluated the use of one or two hidden layers comprising odd numbers of nodes of different topologies: log-sigmoidal and tan-sigmoidal. The output layer was composed of a linear node. The back-propagation algorithm was used to train the neural networks.

We also compared the described classifiers with two heuristic methods frequently used in the literature: Heuristic 1, which considers the singular region nearest to the top of the finger as the core point; and Heuristic 2, which defines the core point as the singular region nearest to the centroid of the ROI.

To train and validate the classifiers, we used the N-fold cross-validation technique with $N = 10$ [331]. Table 6.4 presents the classification accuracy of the evaluated methods over Datasets Ac and Bc.

The results demonstrated that the presented core-detection approach offers remarkable accuracy on all datasets compared with the reference methods. In fact,

Table 6.4 Classifiers Evaluated for Estimating the Core Point

Classifier	Dataset Ac		Dataset Bc	
	Mean	Std	Mean	Std
FFNN-1	0.078	0.032	0.014	0.045
FFNN-3	0.044	0.052	0.057	0.100
FFNN-5	0.048	0.025	0.029	0.060
FFNN-10	0.048	0.025	0.157	0.171
kNN-1	0.013	0.004	0.014	0.000
kNN-3	0.016	0.004	0.028	0.000
kNN-5	0.023	0.002	0.032	0.018
kNN-10	0.028	0.002	0.056	0.000
LDC	0.03	0.002	0.032	0.007
KL-LDC	0.038	0.006	0.039	0.013
PCA-LDC	0.035	0.007	0.028	0.000
Quadratic	0.049	0.006	0.944	0.000
Heuristic 1	0.065	0.248	0.113	0.318
Heuristic 2	0.156	0.364	0.127	0.335

Source: R. Donida Labati, V. Piuri, and F. Scotti, Measurement of the principal singular point in fingerprint images: A neural approach, in *Proceedings of the IEEE International Conference on Computational Intelligence for Measurement Systems and Applications*, pp. 18–23. © 2010, IEEE.

Note: Classification errors obtained for the different datasets using the following methods: a feed-forward neural network with one hidden layer composed of 1 node (FFNN-1), 3 nodes (FFNN-3), 5 nodes (FFNN-5), and 10 nodes (FFNN-10); a k-nearest-neighbor with $k = 1$ (kNN-1), $k = 3$ (kNN-3), $k = 5$ (kNN-5), and $k = 10$ (kNN-10); a normal densities-based linear classifier (ldc); a linear classifier by KL expansion of a common covariance matrix (KL-LDC); a linear classifier by PCA expansion of the joint data (PCA-LDC); a quadratic classifier (Quadratic); Heuristic 1; and Heuristic 2.

for the evaluated datasets, trained classifiers achieved better performance than the heuristic methods considered.

In addition, in classifying Dataset Bc, the classification method based on neural networks exhibited extremely good accuracy with respect to the tested inductive classification techniques.

For the considered datasets, a similar accuracy was obtained only by the kNN classifiers. However, the neural-network approach incurs a relevant gain in the computational complexity (up to a factor of 10 for the experimental datasets). The experiments thus demonstrated that the neural-based quality-classification system is the most suitable for real-time application contexts.

The obtained results suggest that the proposed method is general and can be effectively applied to touch-based and touchless fingerprint images.

6.1.3 *Reduction of Perspective and Rotation Effects*

The studied method for reducing perspective and rotation effects (Section 5.2.6) uses neural networks to estimate the roll angle between two touchless fingerprint samples and then adopts the obtained information to compensate for the orientation difference between the two fingerprint acquisitions by rotating a synthetic three-dimensional model of the finger surface estimated from one of the compared images. The performance of the proposed approach was evaluated using touchless fingerprint images captured with a variety of orientations with respect to the camera. The performed tests examined the proposed method for simulating the finger rotations and the classification accuracy of the proposed technique for estimating the roll angle between two acquisitions. Then, the approach was validated by evaluating the accuracy improvement introduced in a complete biometric system.

6.1.3.1 *Creation of the Evaluation Datasets and Tuning of the Parameters*

The used acquisition setup comprised a Sony XCD-SX90CR CCD color camera and a blue LED with a light diffuser. The setup configuration is presented in Figure 6.4. To control the lens focus, the system required that the finger be placed

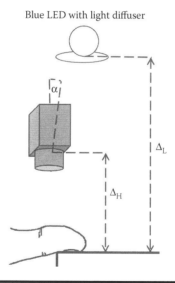

Blue LED with light diffuser

α

Δ_L

Δ_H

Figure 6.4 Schema of the used acquisition setup. During the experiments, the roll angle of the finger was varied by rotating the camera at an angle α along the x-axis. (R. Donida Labati et al., Contactless fingerprint recognition: A neural approach for perspective and rotation effects reduction, in *Proceedings of the IEEE Workshop on Computational Intelligence in Biometrics and Identity Management*, pp. 22–30. © 2013, IEEE.)

on a surface at a known distance from the camera. The setup did not use finger placement guides, thereby allowing the finger to be placed with uncontrolled roll and yaw orientations. The distance from the digital camera sensor to the finger placement surface was $\Delta_H = 215$ mm, and the distance from the LED light array to the placement surface was $\Delta_L = 130$ mm. The captured touchless fingerprint images had a size of 1280×960 pixels.

Because placing the finger with controlled roll rotations with respect to the camera without using guides is likely impossible, we simulated acquisitions with known roll angles by modifying the angle α between the camera and its support to create the following datasets:

- Dataset Ar: 400 touchless samples captured from 50 fingers with $\alpha = -10°$. Each finger was acquired 8 times.
- Dataset Br: 400 touchless samples captured from 50 fingers with $\alpha = +10°$. Each finger was acquired 8 times.
- Dataset Cr: 800 samples pertaining to Dataset Ar and Dataset Br. Sixteen samples were collected for each finger (eight fingerprint images from Dataset Ar and eight fingerprint images from Dataset Br). To properly control the orientation difference between the performed acquisitions and to reduce the effects of micromovements of the finger, samples pertaining to Datasets Ar and Br were acquired at the same instant of time using two cameras synchronized with a trigger mechanism.

The parameters used for simulating synthetic three-dimensional models of the finger surface were $C_H = 40$ and $C_W = 2/5$.

6.1.3.2 Simulation of Finger Rotation

The performance of the studied method for simulating the finger rotation was tested by recovering touchless fingerprint samples captured with different orientations.

The test consisted of a comparison of the matching scores obtained by performing genuine identity comparisons using the baseline biometric recognition technique (Rotation Algorithm 1) and the proposed method (Rotation Algorithm 2). The test was performed on the samples pertaining to Dataset Cr and consisted of 800 genuine identity comparisons.

The compared techniques and evaluation procedures were as follows:

- Rotation Algorithm 1 (no simulated rotations): The NIST MINDTCT and NIST BOZORTH3 software tools were applied after image preprocessing to perform the feature extraction and matching steps.

  The evaluation procedure consisted of a comparison of the templates obtained from touchless images captured at the same instant using different cameras. For each fingerprint image i of Dataset Ar, we performed the following identity comparisons:

$$M(i, 1) = \text{Match}(\text{Dataset Ar}(i), \text{Dataset Br}(i)),$$
$$M(i, 2) = \text{Match}(\text{Dataset Br}(i), \text{Dataset Ar}(i)). \tag{6.1}$$

■ Rotation Algorithm 2 (1 simulated rotation): For each identity comparison, the two considered samples were first treated with the presented preprocessing algorithm. Then, the proposed method for recovering the finger rotation was applied to one of the images using a rotation angle τ equal to the real angular difference $\Delta_\tau = |2\alpha|$. Finally, the feature extraction and matching steps were performed in the same manner as Rotation Algorithm 1.

The evaluation procedure was similar to that used for Rotation Algorithm 1. For each fingerprint image i of Dataset Ar, we performed the following identity comparisons:

$$M(i, 1) = \text{Match}(\text{Rot}(\text{Dataset Ar}(i), \Delta_\theta), \text{Dataset Br}(i)),$$
$$M(i, 2) = \text{Match}(\text{Rot}(\text{Dataset Br}(i), -\Delta_\theta), \text{Dataset Ar}(i)), \tag{6.2}$$

where $\text{Rot}(I, \tau)$ represents the method proposed for simulating touchless acquisitions of rotated fingers.

The performed tests demonstrated that the use of the proposed method for simulating the finger rotation in biometric recognition techniques (Rotation Algorithm 2) permitted a mean matching score increase of 16.7% compared to Rotation Algorithm 1. The results presented in Figure 6.5 demonstrate that the proposed method can increase the matching score between genuine individuals by effectively reducing perspective deformations due to differences in the roll angles of touchless fingerprint acquisitions. The matching scores were also increased in all of the regions of the boxplot.

6.1.3.3 Neural Estimation of the Roll-Angle Difference

We consider the roll-angle estimation as a classification problem in which the orientation difference between two touchless fingerprint acquisitions is estimated as a discrete value representing a range of angles.

The classification performance of the proposed method was evaluated using a feature set related to genuine identity comparisons performed on samples pertaining to Dataset Cr. This feature set comprised 12,000 elements. For each pair of images, the characteristics of the acquisition setup were used to infer a rough measurement of the orientation difference between the touchless fingerprint acquisitions.

Because the acquisitions were performed using two cameras placed at a fixed angular distance, we considered the estimation of the roll-angle difference between two samples as a three-class classification problem. For each possible genuine identity comparison between the touchless fingerprint images i and j of Dataset Cr, the

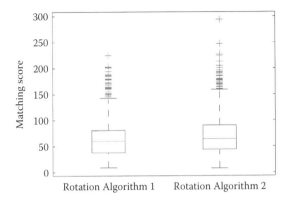

Figure 6.5 Boxplot of the matching scores between genuine samples captured by a difference of 20°. The matching scores were directly computed (Rotation Algorithm 1) by applying the proposed method for simulating the finger rotation (Rotation Algorithm 2). (R. Donida Labati et al., Contactless fingerprint recognition: A neural approach for perspective and rotation effects reduction, in *Proceedings of the IEEE Workshop on Computational Intelligence in Biometrics and Identity Management*, pp. 22–30. © 2013, IEEE.)

class of roll-angle difference was labeled as follows:

$$\Delta_\tau(i,j) = \begin{cases} +20°, & \text{if } (\text{Dataset Cr}(i) \cap \text{Dataset Ar}) \text{ and } (\text{Dataset Cr}(j) \cap \text{Dataset Br}), \\ -20°, & \text{if } (\text{Dataset Cr}(i) \cap \text{Dataset Br}) \text{ and } (\text{Dataset Cr}(j) \cap \text{Dataset Ar}), \\ 0°, & \text{otherwise.} \end{cases}$$

$$(6.3)$$

The difference between the roll angles of two fingerprint images was estimated using classifiers based on feed-forward neural networks. The topology of the neural classifiers was defined as follows: The output layer was composed of a linear node, and we tested different numbers of tan-sigmoidal nodes in a single hidden layer. The generalization capability of the trained classifiers was validated using a two-fold validation approach in which the feature dataset was divided into a training set composed of 6000 randomly selected elements and a validation set comprising the remaining 6000 elements. To estimate the generalization capability of the trained neural networks, we used the two-fold validation strategy and divided the feature dataset into a training set composed of 6000 randomly selected samples and a validation set comprising the remaining 6000 elements.

To better analyze the complexity of the considered learning problem, we compared the performance of the neural classifiers with the results of different classification paradigms. We evaluated the following classifier families:

Table 6.5   Results of Different Classifiers for the Feature Dataset

Classifier	Total Classification Error (%)
Linear	12.17
Quadratic	10.03
kNN-3	3.58
FNN-40	1.65

Source: R. Donida Labati et al., Contactless fingerprint recognition: A neural approach for perspective and rotation effects reduction, in *Proceedings of the IEEE Workshop on Computational Intelligence in Biometrics and Identity Management*, pp. 22–30. © 2013, IEEE.

Note: FNN-40 = Feed-forward neural network with one hidden layer composed of 40 nodes; kNN = k-nearest neighbor, where k stands for the number of first neighbors.

■ The linear classifier
■ The quadratic classifier
■ k-nearest-neighbor classifiers with different values of the parameter k (1, 3, and 5)
■ Feed-forward neural networks with a single hidden layer comprising different numbers of nodes (1, 3, 5, 10, 15, 20, 25, 30, 35, 40, 45, and 50)

Table 6.5 reports the most accurate results obtained using the tested classifiers. Neural networks with a hidden layer composed of 40 nodes clearly yielded the best accuracy for the considered dataset, with a total classification error of 1.65%.

The performance of the best-trained neural classifier is presented in the confusion matrix reported in Table 6.6. This table indicates that most of the errors consisted of falsely estimated angular differences of $0°$. In a biometric system based on the proposed approach, an estimated angular distance equal to $0°$ implies that the identity comparison should be performed considering only the original fingerprint acquisitions. Therefore, falsely estimated angular differences of $0°$ should not decrease the recognition accuracy compared to approaches that do not perform rotation normalizations.

6.1.3.4 Effects of the Proposed Approach on the Performances of the Biometric System

To evaluate the effects of the proposed approach on the performance of a complete biometric recognition system, we compared the following processing configurations using Dataset Cr:

Table 6.6 Confusion Matrix Obtained Using the Best-Trained Classifier

		Predicted (%)	
Actual	$-20°$	$0°$	$+20°$
$-20°$	26.12	0.55	0.00
$0°$	0.29	45.99	0.38
$+20°$	0.00	0.43	26.24

Source: R. Donida Labati et al., Contactless fingerprint recognition: A neural approach for perspective and rotation effects reduction, in *Proceedings of the IEEE Workshop on Computational Intelligence in Biometrics and Identity Management*, pp. 22–30. © 2013, IEEE.

■ Compensation Method A (no simulated rotations): The method did not use the proposed rotation compensation approach and performed the matching step by comparing the minutiae extracted from the enhanced touchless fingerprint images using the software NIST BOZORTH3. In the following, this method is considered as the baseline technique.

■ Compensation Method B (three simulated rotations): The method extracted the minutiae template from the fresh sample in the same manner as Compensation Method A. The fresh template was then compared with the three stored templates rotated by roll angles $\tau = -20°, 0°, +20°$ by applying the software NIST BOZORTH3. The matching score was computed as the highest score obtained by the minutiae matcher on the three performed template comparisons.

■ Compensation Method C (seven simulated rotations): The method is identical to Compensation Method B; however, the considered angles were $\tau = -30°, -20°, -10°, 0°, +10°, +20°, +30°$.

■ Compensation Method D (angular difference estimation via neural classifier): The method computed the fresh template in the same manner as Compensation Method A. Then, the angular difference between the fresh acquisition and the biometric data stored in the database was estimated using the proposed neural classification technique. Finally, the template rotated by the estimated angle was compared with the fresh template by applying the software NIST BOZORTH3. For each identity comparison, this method applied the minutiae matcher once.

The performed test consisted of 12,000 genuine identity comparisons and 627,200 impostor identity comparisons. The obtained DET curves are presented in Figure 6.6, and the EERs of the tested methods are reported in Table 6.7. Notably, the proposed approach (Compensation Method D) achieved a greater accuracy than the baseline approach (Compensation Method A) in all working points

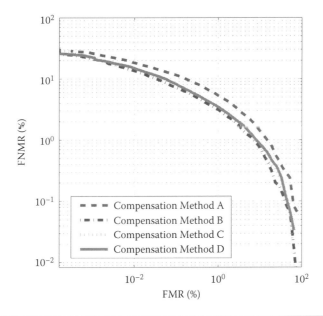

Figure 6.6 **DET curves obtained using the compared biometric recognition methods. Compensation Method A (no rotations), Compensation Method B (three rotations), Compensation Method C (seven rotations), and Compensation Method D (one rotation via Neural Classifier). (R. Donida Labati et al., Contactless fingerprint recognition: A neural approach for perspective and rotation effects reduction, in *Proceedings of the IEEE Workshop on Computational Intelligence in Biometrics and Identity Management*, pp. 22–30. © 2013, IEEE.)**

depicted by the DET curves. Moreover, the EER of Compensation Method D was approximately 1% lower than that of Compensation Method A.

Notably, the proposed approach introduced only a minimal increase in computational complexity in the matching step of the biometric recognition process. For example, the time required to compute the feature set used to estimate the

Table 6.7   EER Values Obtained Using the Compared Methods

Method	EER (%)
Compensation Method A	3.04
Compensation Method B	2.20
Compensation Method C	2.03
Compensation Method D	2.20

Source: R. Donida Labati et al., Contactless fingerprint recognition: A neural approach for perspective and rotation effects reduction, in *Proceedings of the IEEE Workshop on Computational Intelligence in Biometrics and Identity Management*, pp. 22–30. © 2013, IEEE.

roll-angular difference between touchless acquisitions was approximately one-fifth of the required computational time for the BOZORTH3 matcher.

Interestingly, the proposed method for simulating finger rotations applied using sets of previously defined angles achieved similar accuracy to the proposed approach based on neural networks for estimating the roll-angle difference between biometric acquisitions, although with higher computational complexity. For example, Compensation Method B achieved comparable accuracy with respect to Compensation Method D on the considered biometric dataset, but the software BOZORTH3 was applied three times for each identity comparison, thus resulting in an approximately three-fold longer computational time compared with Compensation Method D.

6.2 Methods Based on Three-Dimensional Models

This section describes the tests performed to evaluate the accuracy of the proposed three-dimensional reconstruction techniques and methods for computing touch-equivalent images based on three-dimensional samples. First, the proposed technique for estimating the three-dimensional coordinates of minutiae points (Section 5.3.1) is treated. Then, the three-dimensional reconstruction techniques three-dimensional Method A and three-dimensional Method B (Section 5.3.2) are compared. Finally, the results of the evaluation of the quality of touch-equivalent images obtained by unwrapping three-dimensional fingerprint samples (Sections 5.3.4 and 5.3.5) are discussed.

6.2.1 Three-Dimensional Reconstruction of the Minutiae Points

This section describes the experiments performed to evaluate the proposed technique for the three-dimensional reconstruction of the minutiae points described in Section 5.3.1.

The presented method was written in Matlab language (Version 7.6) on an Intel Centrino 2.0 GHz in Windows XP Professional. The following parameters were used to compute the features related to the FingerCode features: $N_R = 2$, $N_A = 4$, and $N_F = 4$. The following parameters were used to analyze the HOG features: $c_w = 3$, $c_h = 3$, and $c_b = 9$.

6.2.1.1 Creation of the Training and Testing Datasets

We captured in our laboratory a set of 120 color images of two fingers using two Sony SX90CR CCD cameras. The image size was 1280×960 pixels. The schema of the used acquisition setup is described in Section 5.3.1. The illumination was controlled by four white LEDs, and the focal length was 25 mm. We used three different configurations of the acquisition setup, which were created by varying the angle α between the cameras, the distance Δ_D between the CCDs of the cameras, and the distance Δ_H from the CCDs to the surface on which the finger was placed. The setup configurations were implemented as follows:

- Setup 05: $\alpha = 5°$, $\Delta_D = 45$ mm, and $\Delta_H = 230$ mm.
- Setup 10: $\alpha = 10°$, $\Delta_D = 75$ mm, and $\Delta_H = 230$ mm.
- Setup 15: $\alpha = 15°$, $\Delta_D = 125$ mm, and $\Delta_H = 230$ mm.

For each configuration of the acquisition setup, we collected 10 pairs of images for each finger.

Then, a human supervisor searched pairs of corresponding minutiae in different touchless fingerprint images. To simplify this task, the supervisor considered only minutiae present in a circular region radius of 120 pixels around the core of the fingerprint. A total of 24,258 minutiae pairs were labeled as positive (real pairs) or negative (false pairs).

For each configuration of the acquisition setup, we created six feature sets, resulting in 24 feature datasets. For each labeled minutiae pair (i, j) of two touchless samples A and B, these datasets were composed as follows:

- Feature set A: The Euclidean distance between the features $F_A(i)$ and $F_B(j)$.
- Feature set B: The Euclidean distance between the features $H_A(i)$ and $H_B(j)$.
- Feature set C: $M_A(i) - M_B(j)$, and the Euclidean distance between the features $F_A(i)$ and $F_B(j)$.
- Feature set D: $M_A(i) - M_B(j)$, and the Euclidean distance between the features $H_A(i)$ and $H_B(j)$.
- Feature set E: $M_A(i) - M_B(j)$, and $F_A(i) - F_B(j)$.
- Feature set F: $M_A(i) - M_B(j)$, and $H_A(i) - H_B(j)$.

6.2.1.2 Classification of Minutiae Pairs

To evaluate the complexity of the classification problem, we evaluated different families of classifiers:

- The linear Bayes normal classifier (LDC)
- The linear classifier by KL expansion of the common covariance matrix (KLLDC)
- The linear classifier by PCA expansion of the joint data (PCA-LDC)
- The k-nearest-neighbor classifier with different values of the parameter k (1, 3, and 5)
- Feed-forward neural networks with a single hidden layer comprising different numbers of neurons (FFNN)

The neural networks comprised a linear node in the output layer and log-sigmoidal node in the hidden layer. The classical back-propagation algorithm was used as the training method. All considered classifiers were trained and evaluated using the N-fold cross-validation technique with $N = 10$ [331].

The proposed method displayed remarkable accuracy, particularly when neural networks were adopted to classify the minutiae pairs. These experiments demonstrated the suitability of neural classifiers in different setup configurations. Remarkably, neural networks achieved comparable or better accuracy with respect to the most accurate inductive methods among the feature datasets used in these tests. Moreover, neural classifiers presented minor computational complexity.

Table 6.8 presents the results for different feature sets extracted from the touchless fingerprint images captured using Setup 05. The proposed method achieved remarkable accuracy, particularly when features based on Gabor filters (Dataset 05-C) were used, with a classification mean error equal to 0.9%. An analogous situation occurred in the experiments related to Setups 10 and 15. Table 6.9 reports that the neural networks yielded the greatest classification accuracy in Setup 15, with a classification mean error equal to 1.5%. Only the kNN classifiers displayed comparable accuracy for the considered datasets. However, the neural classifiers offered a relevant gain in computational complexity, achieving a minimum gain factor greater than 10.

These experiments demonstrated that the proposed feature sets were capable of classifying minutiae pairs with notable accuracy in touchless fingerprint images and that neural classifiers are likely the most suitable models for real-time application contexts.

6.2.1.3 Computation of Three-Dimensional Minutia

The multiple-view setup was calibrated using 15 different pairs of chessboard images for each configuration of the acquisition setup. The chessboard used comprises 12×9 squares of 2.8×2.8 mm.

The analysis of the obtained three-dimensional minutiae indicated the effectiveness of the three-dimensional reconstruction. The calibration procedure for the system exhibited remarkable three-dimensional reconstruction accuracy with localization errors of less than 0.1 mm. Figure 6.7 presents an example of the reconstructed three-dimensional minutiae. The vertical segments indicate the correspondences between the three-dimensional minutiae and the relative two-dimensional coordinates in the fingerprint image captured by the left camera.

Erroneously classified minutiae pairs usually present three-dimensional coordinates far from the majority of the three-dimensional minutiae. Therefore, classification errors and false minutiae can be easily detected by applying a spike filter.

6.2.2 Three-Dimensional Reconstruction of the Finger Surface

This section compares the performances obtained using three-dimensional Method A and three-dimensional Method B for the proposed approach for

Table 6.8 Results Obtained Using the Proposed Method to Search for Corresponding Minutiae Points Using Neural Networks on the 05° Datasets

Method	Dataset 05-A		Dataset 05-B		Dataset 05-C		Dataset 05-D		Dataset 05-E		Dataset 05-F	
	Mean	Std	Mean	Std	Mean	Std	Mean	Std	Mean	Std	Mean	Std
Linear	0.010	0.000	0.017	0.000	0.009	0.001	0.015	0.000	0.037	0.000	0.038	0.000
Quadratic	0.010	0.000	0.018	0.000	0.008	0.000	0.008	0.000	0.012	0.001	0.036	0.001
pcldc	0.010	0.000	0.017	0.000	0.009	0.000	0.015	0.001	0.037	0.000	0.038	0.001
klldc	0.010	0.000	0.017	0.000	0.009	0.000	0.015	0.000	0.037	0.000	0.038	0.000
kNN-1	0.018	0.001	0.031	0.001	0.011	0.001	0.016	0.001	0.013	0.001	0.016	0.001
kNN-3	0.012	0.000	0.022	0.001	0.010	0.001	0.014	0.001	0.014	0.001	0.015	0.001
kNN-5	0.011	0.001	0.021	0.001	0.010	0.000	0.013	0.001	0.015	0.000	0.014	0.000
kNN-10	0.011	0.001	0.020	0.001	0.009	0.001	0.015	0.000	0.020	0.000	0.015	0.000
FFNN-1	0.015	0.016	0.026	0.015	0.008	0.004	0.014	0.006	0.040	0.016	0.065	0.020
FFNN-3	0.010	0.006	0.018	0.010	0.007	0.003	0.015	0.010	0.049	0.016	0.063	0.014
FFNN-5	0.0100	0.005	0.019	0.007	0.010	0.003	0.013	0.008	0.042	0.025	0.056	0.026
FFNN-10	0.010	0.006	0.020	0.009	0.014	0.012	0.008	0.006	0.030	0.018	0.049	0.020

Source: R. Donida Labati, V. Piuri, and F. Scotti, A neural-based minutiae pair identification method for touchless fingerprint images, in *Proceedings of the IEEE Workshop on Computational Intelligence in Biometrics and Identity Management,* © 2011, IEEE.

Note: Classification methods: a linear classifier (linear); a quadratic classifier (quadratic); a linear classifier using PC expansion (pcldc); a linear classifier using KL expansion (klldc); a kNN with $k = 1$ (kNN-1); a kNN with $k = 3$ (kNN-3); a kNN with $k = 5$ (kNN-5); a kNN with $k = 10$ (kNN-10); a feed-forward neural network with one hidden layer composed of 1 node (FFNN-1); a feed-forward neural network with one hidden layer composed of 3 nodes (FFNN-3); a feed-forward neural network with one hidden layer composed of 5 nodes (FFNN-5); and a feed-forward neural network with one hidden layer composed of 10 nodes (FFNN-10).

Table 6.9 Results Obtained Using the Proposed Method to Search for Corresponding Minutiae Points Using Neural Networks on the 15° Datasets

Method	Dataset 15-A		Dataset 15-B		Dataset 15-C		Dataset 15-D		Dataset 15-E		Dataset 15-F	
	Mean	Std	Mean	Std	Mean	Std	Mean	Std	Mean	Std	Mean	Std
Linear	0.016	0.001	0.018	0.000	0.020	0.000	0.018	0.001	0.044	0.001	0.052	0.002
Quadratic	0.020	0.000	0.017	0.000	0.018	0.000	0.015	0.001	0.040	0.001	0.146	0.007
pcldc	0.016	0.001	0.018	0.001	0.020	0.000	0.017	0.001	0.043	0.000	0.052	0.002
klldc	0.016	0.001	0.017	0.000	0.020	0.000	0.018	0.001	0.044	0.000	0.052	0.002
kNN-1	0.023	0.001	0.031	0.002	0.027	0.002	0.020	0.002	0.017	0.000	0.019	0.001
kNN-3	0.021	0.002	0.020	0.002	0.021	0.001	0.017	0.001	0.022	0.002	0.020	0.002
kNN-5	0.018	0.002	0.017	0.001	0.015	0.001	0.020	0.001	0.027	0.003	0.019	0.002
kNN-10	0.014	0.000	0.015	0.001	0.015	0.001	0.019	0.000	0.030	0.002	0.018	0.001
FFNN-1	0.024	0.020	0.024	0.022	0.020	0.015	0.020	0.017	0.059	0.024	0.065	0.015
FFNN-3	0.016	0.013	0.016	0.009	0.019	0.009	0.022	0.018	0.067	0.022	0.069	0.029
FFNN-5	0.015	0.012	0.017	0.009	0.015	0.012	0.018	0.011	0.078	0.037	0.050	0.017
FFNN-10	0.015	0.014	0.019	0.017	0.026	0.015	0.023	0.023	0.044	0.034	0.096	0.126

Source: R. Donida Labati, V. Piuri, and F. Scotti, A neural-based minutiae pair identification method for touchless fingerprint images, in *Proceedings of the IEEE Workshop on Computational Intelligence in Biometrics and Identity Management*, © 2011, IEEE.

Note: Classification methods: a linear classifier (linear); a quadratic classifier (quadratic); a linear classifier using KL expansion (klldc); a linear classifier using PC expansion (pcldc); a kNN with $k = 1$ (kNN-1); a kNN with $k = 3$ (kNN-3); a kNN with $k = 5$ (kNN-5); a kNN with $k = 10$ (kNN-10); a feed-forward neural network with one hidden layer composed of 1 node (FFNN-1); a feed-forward neural network with one hidden layer composed of 3 nodes (FFNN-3); a feed-forward neural network with one hidden layer composed of 5 nodes (FFNN-5); and a feed-forward neural network with one hidden layer composed of 10 nodes (FFNN-10).

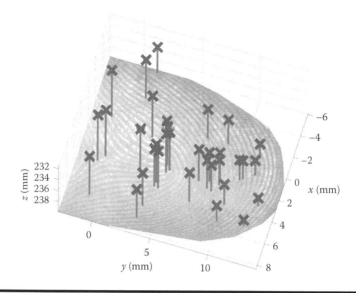

Figure 6.7 **Example of three-dimensional minutiae reconstructed using the proposed method for identifying corresponding minutiae points using neural networks. The vertical segments indicate the correspondence between the identified three-dimensional points and the relative positions of the minutiae of the left image. The distances are related to the reference viewpoint and are expressed in mm. (R. Donida Labati, V. Piuri, and F. Scotti, A neural-based minutiae pair identification method for touchless fingerprint images, in** *Proceedings of the IEEE Workshop on Computational Intelligence in Biometrics and Identity Management,* © **2011, IEEE.)**

reconstructing the finger volume (Section 5.3.2). This comparison is first performed by evaluating the quality of the touch-equivalent images obtained by the two methods.

6.2.2.1 The Used Datasets

The performances of three-dimensional Method A and three-dimensional Method B were compared using biometric samples captured in our laboratory. The acquisition setups used in the evaluated three-dimensional reconstruction methods are presented in Figure 5.14. The parameters of the acquisition setup were as follows: the angle α between the cameras and their horizontal support was $85°$, the distance Δ_D between the centers of the CCDs was 45 mm, the distance Δ_H between the CCDs and the surface used to place the finger was 235 mm, and the distance Δ_L between the LED light and the surface was 290 mm.

We created two image datasets to evaluate the performance of three-dimensional Method A and three-dimensional Method B. Each dataset comprised 360 pairs of images captured from 36 fingers using the two-view acquisition system. For each finger, we performed 10 acquisitions. We referred to the dataset captured using the acquisition setup of three-dimensional Method A as Dataset A3d and to the dataset captured using three-dimensional Method B as Dataset B3d.

6.2.2.2 Parameters of the Proposed Methods

We empirically tuned the parameters of the proposed methods for the datasets employed. The value of Δ_P used to compute the squares of the project pattern was 80. The values used to search the corresponding pairs of points in the two-view images were $l = 21$, and $w = 130$. The parameters used to remove outliers from the three-dimensional models were $t_{s1} = 8$, $t_{s2} = 7$ and the region S_A is composed by the nearest eight points. The values used to compute the touch-equivalent images were $t_x = 8$ and $t_z = 8$.

6.2.2.3 Accuracy of the Three-Dimensional Reconstruction Methods

First, we computed the calibration error of the three-dimensional reconstruction system. The acquisition setup was calibrated using the same method adopted to calibrate the setup described in Section 6.2.1. We obtained a reconstruction error of 0.03 mm. This error was computed using the technique describe in [335]. First, we computed the three-dimensional coordinates of the corners detected in the calibration images. Then, a plane interpolating the reconstructed three-dimensional points was computed. Finally, the calibration error was computed as the standard deviation of the Euclidean distance between the three-dimensional corners and the interpolating plane.

Three-dimensional models were computed from the images pertaining to the considered datasets using three-dimensional Method A and three-dimensional Method B, and the quality of the resulting unwrapped three-dimensional models was evaluated. One important quality measure is the evaluation of the ridge pattern visibility and distortion after the unwrapping step. Low-quality ridge patterns can drastically reduce the accuracy of a biometric recognition system. This analysis is particularly important in the proposed system because an incorrect projected pattern would drastically decrease the visibility of the ridges. Figure 6.8 shows two examples of enhanced fingertip images after the unwrapping of the three-dimensional models computed using three-dimensional Method A and three-dimensional Method B. The images reveal that the use of structured light does not affect the quality of the resulting textures.

Another important test is the evaluation of the number of spikes in the reconstructed three-dimensional models. These points can introduce distortions and artifacts in the three-dimensional finger shape and in the final touch-equivalent image. The number of spikes obtained using three-dimensional Method A was

(a)   (b)

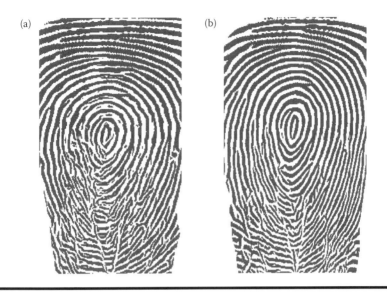

Figure 6.8 Touch-equivalent images obtained by unwrapping the computed three-dimensional models using structured-light and nonstructured light approaches: (a) an image obtained using three-dimensional Method A and (b) an image obtained using three-dimensional Method B. The use of structured light can accelerate the search of corresponding points in two-view images by up to 90%. (R. Donida Labati et al., Fast 3-D fingertip reconstruction using a single two-view structured light acquisition, in *Proceedings of the IEEE Workshop on Biometric Measurements and Systems for Security and Medical Applications*, pp. 1–8. © 2011, IEEE.)

smaller than that obtained using three-dimensional Method B. Therefore, three-dimensional reconstructions that are more accurate can be obtained using simpler and faster filtering techniques. Figure 6.9 presents a comparison of the filtered and unfiltered point clouds and the respective surface estimations. Notably, the small number of outliers did not affect the quality of the reconstructed three-dimensional surface.

To numerically assess the quality of the reconstructed three-dimensional models, we analyzed the computed touch-equivalent images using the software NIST NFIQ. This test was performed because the quality of the touch-equivalent images can reveal the presence of deformations and outliers in the corresponding three-dimensional model. Table 6.10 presents the corresponding results. This table indicates that processing fingertip models using three-dimensional Method A yields touch-equivalent images of better quality due to the influence of spikes on the quality of the images obtained.

We also compared the computational time required for the two three-dimensional reconstruction methods. The use of the projected pattern enabled a dramatic reduction in the time required for point matching of approximately 90%.

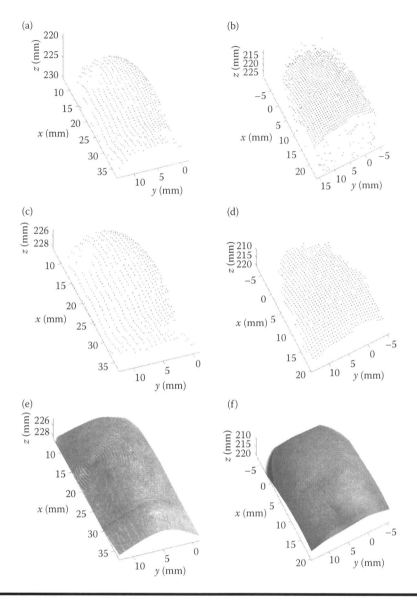

Figure 6.9 Three-dimensional point clouds, filtered and unfiltered, computed using three-dimensional Method A and three-dimensional Method B: (a, b) filtered point cloud and surface mapping using three-dimensional Method A; (c, d) unfiltered point cloud and surface mapping using three-dimensional Method A; (e, f) filtered point cloud and surface mapping using three-dimensional Method B; and unfiltered point cloud and surface mapping using three-dimensional Method B. (R. Donida Labati et al., Fast 3-D fingertip reconstruction using a single two-view structured light acquisition, in *Proceedings of the IEEE Workshop on Biometric Measurements and Systems for Security and Medical Applications*, pp. 1–8. © 2011, IEEE.)

Table 6.10 Numerical Evaluation of the Unwrapped Fingerprint Images Obtained from Three-Dimensional Samples Representing the Finger Volume

Three-Dimensional Method A		Three-Dimensional Method B	
Mean	*Std*	*Mean*	*Std*
1.380	0.840	1.890	1.120

Source: R. Donida Labati et al., Fast 3-D fingertip reconstruction using a single two-view structured light acquisition, in *Proceedings of the IEEE Workshop on Biometric Measurements and Systems for Security and Medical Applications*, pp. 1–8. © 2011, IEEE.

6.2.3 *Quality Assessment of Unwrapped Fingerprint Images*

The proposed approach for estimating the quality of touch-equivalent fingerprint images obtained by unwrapping three-dimensional models (Section 5.3.5) was evaluated using a set of 300 images captured from 30 individuals (10 images per individual). These images had a resolution of approximately 500 ppi and were computed using three-dimensional Method A, which is described in Section 5.3.2, and the unwrapping method described in Section 5.3.4. The configuration of the acquisition setup and the calibration technique were the same as those described in Section 6.2.2. The pairs of touchless fingerprint images used to compute the evaluation set of touch-equivalent images were affected by the following types of problems: reflections, lack of focus, and dirty fingers. Due to these problems, three-dimensional models computed from some pairs of touchless images presented distortions, spikes, and regions with low visibility of the ridge pattern. Therefore, the corresponding touch-equivalent images presented deformations and artifacts.

We automatically labeled each touch-equivalent fingerprint image with a quality value. Similar to the approach described in [217], we considered each label a predictor of a matcher's performance. The first step of the labeling process was the computation of the genuine and impostor distributions of the matching scores obtained by applying the software NIST BOZORTH3 [186] to the minutiae templates extracted from the considered touch-equivalent images. Then, we computed a metric called the normalized matching score to express the capability of a sample to be properly matched with other samples of the same individual. This measure evaluates whether the biometric sample contains sufficient information and is defined as follows:

$$o(x_i) = s_m(x_{ii}) + \frac{\mu(s_m(x_{ij})) - s_m(x_{ii})}{\sigma(s_m(x_{ij}))}, \tag{6.4}$$

where x_i is the considered sample, $s_m(x_{ii})$ is the matching score obtained by comparing x_i with itself, and $\mu(s_m(x_{ij}))$ and $\sigma(s_m(x_{ij}))$ are the mean and standard deviation, respectively, of the matching scores computed by comparing x_i with all other samples from the same individual. To maximize the distance between the genuine and

(a)   (b)   (c)

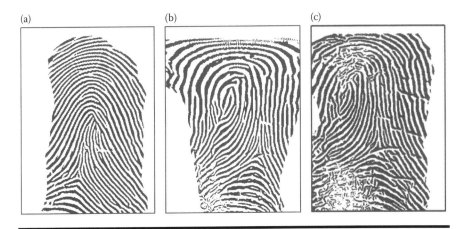

Figure 6.10 Examples of classified touch-equivalent fingerprint images: (a) a sufficient-quality image, (b) a poor-quality image with shape deformations, and (c) a poor-quality image affected by the presence of artifacts. (R. Donida Labati et al., Quality measurement of unwrapped three-dimensional fingerprints: A neural networks approach, in *Proceedings of the International Joint Conference on Neural Networks*, pp. 1123–1130. © 2012, IEEE.)

impostor distributions of the matching scores, we labeled each touch-equivalent image x_i as follows:

$$q(x_i) = \begin{cases} +1 & \text{if } o(x_i) > 96, \\ -1 & \text{otherwise.} \end{cases} \qquad (6.5)$$

The value $+1$ was assigned to the positive class "sufficient" and -1 was assigned to the negative class "poor." In total, 244 "sufficient" and 56 "poor" images were obtained.

Figure 6.10 presents examples of touch-equivalent fingerprint images of different quality. The touch-equivalent fingerprint in Figure 6.10a presents sufficient quality. The image in Figure 6.10b is considered poor because it suffers from deformations of the ridge pattern caused by errors in the reconstruction of the three-dimensional finger shape. Figure 6.10c presents another poor-quality image, which is affected by artifacts caused by out-of-focus regions in the corresponding pair of touchless images.

6.2.4 Classification Results

Different configurations of the feature-extraction algorithm were experimentally evaluated. We used three different sets of angles θ to compute the Gabor features, which were defined as follows:

- $\Theta_a = (0°, 90°)$

- $\Theta_b = (-45°, 45°)$
- $\Theta_c = (-45°, 0°, 45°, 90°)$

We also considered values of the parameters m_G and n_G of 1 to 6. Similarly, we evaluated values of 1–6 for the parameters c_w and c_h, which determine the number of local areas in which the input image is divided during the extraction of the HOG features, and values of 3–12 for the parameter c_b, which corresponds to the number of orientation bins used.

Different combinations of features were used to create 24 feature sets, which are reported in Table 6.11.

The considered configurations of the feature set were experimentally evaluated using classification methods based on feed-forward neural networks. The neural

Table 6.11 Feature Datasets Used to Evaluate the Proposed Approach for the Quality Assessment of Touch-Equivalent Fingerprint Images

Feature Set Name	Minutiae	ROI	F_G, Θ_a	F_G, Θ_b	F_G, Θ_c	F_H	$F_{G\sigma}, \Theta_c$	$F_{H\sigma}$
Gabor-a1			✓					
Gabor-a2	✓		✓					
Gabor-a3		✓	✓					
Gabor-a4	✓	✓	✓					
Gabor-b1				✓				
Gabor-b2	✓			✓				
Gabor-b3		✓		✓				
Gabor-b4	✓	✓		✓				
Gabor-c1					✓			
Gabor-c2	✓				✓			
Gabor-c3		✓			✓			
Gabor-c4	✓	✓			✓			
HOG-1						✓		
HOG-2	✓					✓		
HOG-3		✓				✓		
HOG-4	✓	✓				✓		
Gabor-std-1							✓	
Gabor-std-2	✓						✓	
Gabor-std-3		✓					✓	
Gabor-std-4	✓	✓					✓	
HOG-std-1								✓
HOG-std-2	✓							✓
HOG-std-3		✓						✓
HOG-std-4	✓	✓						✓

Source: R. Donida Labati et al., Quality measurement of unwrapped three-dimensional fingerprints: A neural networks approach, in *Proceedings of the International Joint Conference on Neural Networks*, pp. 1123–1130. © 2012, IEEE.

networks comprised an output node consisting of a linear node and a single hidden layer consisting of tan-sigmoidal nodes. Different configurations of the hidden layer were evaluated by varying the number of its nodes. The back-propagation algorithm was used to train the neural networks. To properly validate the trained neural networks, we adopted the N-fold cross-validation method with $N = 10$ [331].

Table 6.12 reports the results obtained by applying the neural classifiers to the evaluated configurations of the feature set. This table only presents the results of the neural classifiers that achieved the best results and reports the parameters of the feature extraction algorithms, the number of values composing the feature sets, and the composition of the hidden layer. The results obtained indicate that the proposed method achieved remarkable accuracy for all considered configurations of the feature set. Moreover, the best configurations yielded a total classification error equal to 1%. The features based on Gabor filters (F_G and $F_{G\sigma}$) were clearly the most discriminating features. Interestingly, the accuracy achieved using the features F_G and F_H was extremely similar to that obtained using $F_{G\sigma}$ and $F_{H\sigma}$.

Using the best feature set configuration (Gabor-std-1 with $m_G = 6$, $n_G = 6$, and $\theta = 4$), we compared the accuracy of the neural classifiers with the results obtained for the following families of classifiers:

■ Linear classifier
■ Quadratic classifier
■ k-nearest-neighbor with different values of the parameter k

The results obtained are presented in Table 6.13 and indicated that simple classifiers do not yield satisfactory accuracy for the evaluated dataset. In contrast, classifiers that approximate functions that are more complex, such as neural networks with many hidden neurons, can significantly reduce the classification error. For example, neural networks with 45 hidden nodes achieved a 10-fold lower classification error compared to simpler classifiers.

6.2.5 Comparison with Literature Methods

The distribution of the quality classes estimated by the reference software NIST NFIQ [186,221] for the considered dataset of touch-equivalent images is presented in Figure 6.11. NFIQ software is designed for the quality assessment of touch-based fingerprint samples, and it produced sufficient results for the evaluation of touch-equivalent images computed from three-dimensional models reconstructed from touchless acquisitions.

To evaluate the impact of the two methods on the performance of a complete biometric recognition system, we used the considered quality assessment techniques to discard touch-equivalent fingerprint images of poor quality and then tested the verification accuracy of the system using images of sufficient quality. The results

Table 6.12 Accuracy of Neural Classifiers for the Quality Estimation of Touch-Equivalent Fingerprint Images Using Different Feature Sets

Feature Set	Parameters	Feat. #	Hidden	TP (%)	FN (%)	FP (%)	TN (%)	Tot. (%)
Gabor-a1	$m_G=3, n_G=3, \theta_G=2$	18	60	18.00	0.67	2.33	79.00	3.00
Gabor-a2	$m_G=5, n_G=5, \theta_G=2$	55	35	18.00	0.67	0.67	80.67	1.33
Gabor-a3	$m_G=4, n_G=4, \theta_G=2$	36	45	17.67	1.00	1.00	80.33	2.00
Gabor-a4	$m_G=4, n_G=4, \theta_G=2$	39	75	17.67	1.00	1.00	80.33	2.00
Gabor-b1	$m_G=4, n_G=4, \theta_G=2$	32	70	17.00	1.67	0.67	80.67	2.33
Gabor-b2	$m_G=5, n_G=5, \theta_G=2$	55	55	17.67	1.00	0.33	81.00	1.33
Gabor-b3	$m_G=4, n_G=4, \theta_G=2$	36	70	17.67	1.00	2.00	79.33	3.00
Gabor-b4	$m_G=5, n_G=5, \theta_G=2$	57	75	17.33	1.33	1.00	80.33	2.33
Gabor-c1	$m_G=5, n_G=3, \theta_G=4$	36	50	16.67	2.00	0.33	81.00	2.33
Gabor-c2	$m_G=5, n_G=5, \theta_G=4$	67	75	18.33	0.33	0.67	80.67	1.00
Gabor-c3	$m_G=4, n_G=4, \theta_G=4$	68	55	16.33	2.33	0.67	80.67	3.00
Gabor-c4	$m_G=4, n_G=4, \theta_G=4$	71	30	17.67	1.00	0.33	81.00	1.33
HOG-1	$c_w=3, c_h=3, c_b=12$	108	55	16.67	2.00	1.00	80.33	3.00
HOG-2	$c_w=3, c_h=3, c_b=9$	86	35	17.00	1.67	0.33	81.00	2.00
HOG-3	$c_w=3, c_h=3, c_b=9$	85	55	16.00	2.67	0.33	81.00	3.00
HOG-4	$c_w=3, c_h=3, c_b=9$	88	50	17.33	1.33	1.00	80.33	2.33
Gabor-std-1	$m_G=6, n_G=6, \theta_G=4$	36	45	18.00	0.67	0.33	81.00	1.00
Gabor-std-2	$m_G=4, n_G=4, \theta_G=4$	20	65	17.00	1.67	0.33	81.00	2.00
Gabor-std-3	$m_G=6, n_G=6, \theta_G=4$	39	40	17.33	1.33	0.33	81.00	1.67
Gabor-std-4	$m_G=5, n_G=5, \theta_G=4$	32	60	17.00	1.67	0.00	81.33	1.67
HOG-std-1	$c_w=3, c_h=3, c_b=9$	9	55	14.67	4.00	0.33	81.00	4.33
HOG-std-2	$c_w=3, c_h=3, c_b=9$	13	55	17.33	1.33	2.00	79.33	3.33
HOG-std-3	$c_w=3, c_h=3, c_b=9$	12	80	18.00	0.67	2.00	79.33	2.67
HOG-std-4	$c_w=3, c_h=3, c_b=9$	16	70	16.33	2.33	1.33	80.00	3.67

Source: R. Donida Labati et al., Quality measurement of unwrapped three-dimensional fingerprints: A neural networks approach, in *Proceedings of the International Joint Conference on Neural Networks*, pp. 1123–1130. © 2012, IEEE.

Note: Feat. # = number of features; Hidden = number of hidden layer nodes of the feed-forward neural networks; TP = true positives; FN = false-negatives; FP = false-positives; TN = true negatives; Tot. = total classification error.

Table 6.13   Results of Different Classifiers of the Best Feature Set (Gabor-std-1) for Classifying Touch-Equivalent Fingerprint Images

Classifier	Total	Std
FFNN-45	0.01	0.04
ldc	0.18	0.013
klldc	0.18	0.011
pcldc	0.18	0.006
quad	0.17	0.002
kNN-1	0.16	0.007
kNN-3	0.16	0.003
kNN-5	0.15	0.007
kNN-10	0.16	0.002

Source: R. Donida Labati et al., Quality measurement of unwrapped three-dimensional fingerprints: A neural networks approach, in *Proceedings of the International Joint Conference on Neural Networks*, pp. 1123–1130. © 2012, IEEE.

Note: Total = total classification error; Std = standard deviation of the classification error; FFNN-45 = feed-forward neural networks with one hidden layer composed of 45 nodes; lin = linear classifier; klldc = linear classifier using KL expansion; pcldc = linear classifier using PC expansion; quad = quadratic classifier; kNN = k-nearest-neighbor, where k indicates the number of first neighbors.

obtained by applying the minutiae matching software NIST BOZORTH3 [186] are presented in Table 6.14 and Figure 6.12.

Table 6.14 and Figure 6.12 indicate that the use of the proposed approach for discarding poor-quality images effectively increased the recognition accuracy of the biometric system. For example, the EER achieved without using the proposed quality assessment approach was 9.56%, and the EER achieved by discarding the touch-equivalent images classified as poor-quality samples by the proposed approach was 1.97%. Moreover, the proposed approach achieved better results than the software NIST NFIQ. In fact, the proposed approach discarded a smaller number of touch-equivalent fingerprint images (56 vs. 62) and achieved a better EER (1.97% vs. 5.76%). Moreover, these experiments demonstrated that the recognition accuracy was enhanced for nearly all of the DET curve plot compared with the reference methods. This enhancement is attributable to the differences in the types of problems that can affect fingerprint images captured by touch-based sensors and those obtained by unwrapping three-dimensional finger models. Notably, the lower overall accuracy of the biometric system compared to state-of-the-art systems based on touch-based sensors may be due to the particularly noisy fingerprint images that we used to evaluate the proposed approach.

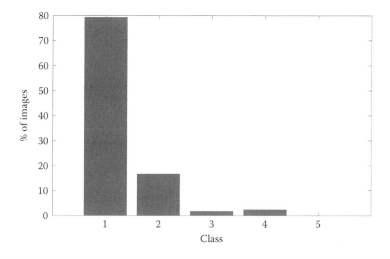

Figure 6.11 Class distribution of the reference software NIST NFIQ for the studied dataset of touch-equivalent fingerprint images. (R. Donida Labati et al., Quality measurement of unwrapped three-dimensional fingerprints: A neural networks approach, in *Proceedings of the International Joint Conference on Neural Networks*, pp. 1123–1130. © 2012, IEEE.)

Table 6.14   Effects on the EER of the Application of Different Quality Classifiers for the Studied Dataset of Touch-Equivalent Fingerprint Images

Data	*Retained Samples*	*Discarded Samples*	*EER (%)*
Original dataset	300	0	9.56
$q_{NFIQ} = 1$	238	62	5.76
$q_{NFIQ} \leq 2$	288	12	7.45
Proposed method	244	56	1.97

Source: R. Donida Labati et al., Quality measurement of unwrapped three-dimensional fingerprints: A neural networks approach, in *Proceedings of the International Joint Conference on Neural Networks*, pp. 1123–1130. © 2012, IEEE.

6.3 Comparison of Biometric Recognition Methods

We performed a scenario evaluation to compare the proposed touchless fingerprint recognition techniques with traditional biometric systems that require touch-based acquisitions.

The performed comparison considers all of the evaluation aspects described in Section 2.3.2: (I) accuracy, (II) speed, (III) cost, (IV) scalability, (V) interoperability, (VI) usability, (VII) social acceptance, (VIII) security, and (IX) privacy.

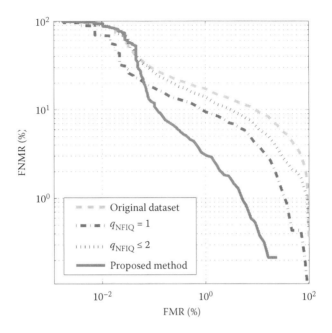

Figure 6.12 Effects on the DET curves of the application of the proposed quality assessment method and NIST NFIQ software to the test dataset composed of 300 touch-equivalent images of three-dimensional fingerprint models. (R. Donida Labati et al., Quality measurement of unwrapped three-dimensional fingerprints: A neural networks approach, in *Proceedings of the International Joint Conference on Neural Networks*, pp. 1123–1130. © 2012, IEEE.)

First, the datasets and application conditions used are described. Then, the parameters of the proposed methods are reported. Finally, the evaluated aspects of the considered biometric systems are analyzed.

6.3.1 The Used Datasets

A scenario evaluation was performed by analyzing the performances of different biometric technologies for access control in a laboratory.

We captured two datasets of fingerprint samples using touchless and touch-based sensors. The samples pertaining to Dataset C3d were acquired using three-dimensional Method C, which is described in Section 5.3.2. The fingerprint images pertaining to Dataset Ct were acquired using CrossMatch V300 [332,333].

The samples pertaining to Dataset C3d and Dataset Ct were captured during a single 1-week session. Dataset C3d was composed of 1040 synchronously captured pairs of touchless fingerprint images, and Dataset Ct was composed of 1040 touch-based fingerprint images. These datasets contained samples captured from the same fingers. Biometric data were captured from a set of 13 volunteers that included both men and women ranging in age from 24 to 63 years and included graduate students, workers, a pensioner, and so on. For both databases, each volunteer contributed 80 acquisitions of the 10 fingers (left and right hands). Thus, each finger was acquired eight times.

The acquisition setup used to create Dataset C3d was composed of two synchronized Sony XCD-SX90CR CCD color cameras and a blue LED with a diffuser lens (Section 6.2.2). The following setup parameters were used: $\alpha = 85°$, $\Delta_D = 45$ mm, $\Delta_L = 90$ mm and $\Delta_H = 240$ mm. The two-view system was calibrated using the technique described in [322,323] with 12 pairs of chessboard images captured in different positions. The calibration chessboard was composed of 12×9 squares of 2.8×2.8 mm.

6.3.2 Parameters Used by Touchless Techniques

The parameters of the proposed methods were empirically tuned for the used dataset. The values used for three-dimensional Method C for matching the reference points were $l = 21$ and $w = 70$; and the values used for the unwrapping technique were $t_x = 8$ and $t_z = 8$.

6.3.3 Accuracy

The accuracy of the proposed touchless fingerprint recognition techniques based on two-dimensional and three-dimensional samples was compared with that achieved by traditional biometric systems using touch-based acquisition methods.

First, we present the results obtained using the proposed touchless recognition methods based on two-dimensional and three-dimensional samples. Then, the best-obtained results are compared with those results achieved using a touch-based recognition system based on well-known algorithms described in the literature.

6.3.3.1 Results of the Approach Based on Two-Dimensional Samples

The accuracy of the proposed biometric recognition technique based on the analysis of Level 2 features of touchless two-dimensional samples (Section 5.2) was evaluated. This method computes touch-equivalent images from fingerprint acquisitions obtained using single cameras and then performs feature extraction and matching using well-known algorithms designed for touch-based fingerprint images.

This method was tested on images captured by the single views of Dataset C3d. Then, this dataset was divided into Dataset C3d-1 and Dataset C3d-2. The first subset contained the images captured by camera A, and the second subset contained the images captured by camera B. Examples of touch-equivalent images related to

(a)   (b)

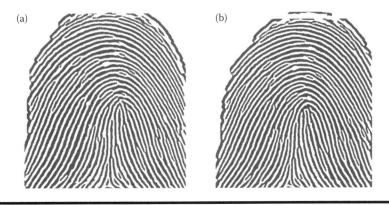

Figure 6.13 Examples of touch-equivalent images of the two views of the proposed acquisition system: (a) camera A and (b) camera B.

the two views of the acquisition system are presented in Figure 6.13. These images clearly present differences due to perspective distortions.

The DET curves obtained by the minutiae-based recognition technique on Dataset C3d-1 and Dataset C3d-2 are presented in Figure 6.14. The obtained EER values on the evaluated datasets are 3.52% and 1.32%, respectively. These results indicate that the level of accuracy achieved using the proposed technique based on single cameras is sufficient for low-cost applications. For example, this technique can be used in mobile devices with integrated cameras.

Another interesting result is that the proposed method achieved better results for Dataset C3d-2 than Dataset C3d-1, most likely due to perspective deformations of the images captured by camera A.

We also evaluated a multimodal biometric system that fuses the matching scores obtained by applying the proposed approach based on touchless two-dimensional samples to images captured by the two views of the considered acquisition setup. This technique can be considered a multimodal system based on multiple snapshots that performs the fusion at the matching score level [336]. We evaluated different fusion functions; however, we did not consider data normalization techniques because the matching values were obtained using the same recognition algorithm with similar data. Considering the matching scores $m_A(i,j)$ and $m_B(i,j)$, the strategies that can be adopted for computing the final matching score $m_s(i,j)$ are

$$m_s(i,j) = \text{mean}(m_A(i,j), m_B(i,j)),$$

$$m_s(i,j) = \min(m_A(i,j), m_B(i,j)), \qquad (6.6)$$

$$m_s(i,j) = \max(m_A(i,j), m_B(i,j)).$$

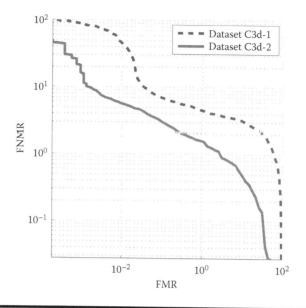

Figure 6.14 DET curves obtained by applying the proposed recognition technique on two-dimensional samples from Dataset C3d-1 and Dataset C3d-2.

The best results are obtained by using the mean fusion function, and achieved EER equal to 0.90%. The DET curve for the multimodal technique with Dataset C3d is presented in Figure 6.15, and the FMR and FNMR at different points of the DET curve are reported in Table 6.15.

The presented multimodal technique effectively increased the recognition accuracy compared the technique based on single touchless acquisitions. Moreover, the obtained results are comparable with touch-based fingerprint recognition systems described in the literature. Table 6.15 also indicates that this system yields good results with thresholds of the matching scores that produce small numbers of false matches. However, the performances decreased drastically with threshold values that produced small numbers of false non-matches.

6.3.3.2 Results of the Approach Based on Three-Dimensional Samples

The accuracy obtained by applying the matching algorithm NIST BOZORTH3 [186] to the touch-equivalent images obtained by applying proposed three-dimensional Method C to Dataset C3d was evaluated. Figure 6.16 presents an example of a three-dimensional fingerprint model and its corresponding touch-equivalent image.

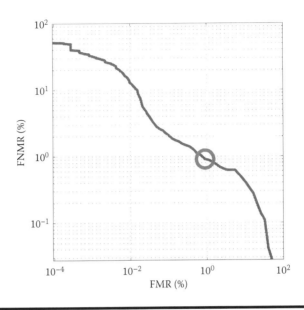

Figure 6.15 DET curve obtained using the proposed multimodal approach based on multiple two-dimensional samples for Dataset C3d. The EER position is marked by a circle in the plot (EER = 0.90%).

Figure 6.17 presents the DET curve obtained by performing 1,080,560 identity comparisons on Dataset C3d, and Table 6.16 depicts the obtained FMR and FNMR values at different points of the DET curve.

The obtained results indicate that the proposed approach can obtain accurate results. In fact, the obtained EER was 0.31%. Moreover, the proposed approach obtained good performances with thresholds of the matching scores that produced a small number of false matches. This characteristic will enable the application of this approach in high-security applications. However, the recognition accuracy was less satisfactory, with thresholds corresponding to small numbers of false non-matches, possibly due to poor-quality three-dimensional reconstructions due to the presence of motion blur in the captured touchless images.

Table 6.15 FMR and FNMR Values Obtained Using the Proposed Multimodal Approach Based on Multiple Two-Dimensional Samples for Dataset C3d

EER (%)	FNMR (%) at FMR = 0.01%	FNMR (%) at FMR = 0.10%
0.90	12.71	2.06

(a)   (b)

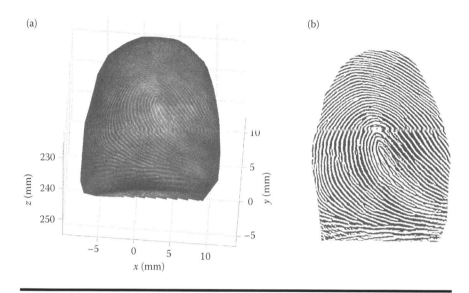

Figure 6.16 **Examples of results obtained using three-dimensional Method C: (a) a three-dimensional fingerprint model and (b) its corresponding touch-equivalent image.**

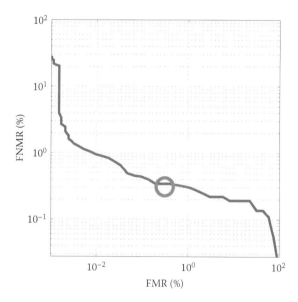

Figure 6.17 **DET curve obtained using recognition techniques designed for touch-based recognition systems on touch-equivalent images computed using proposed three-dimensional Method C and the unwrapping technique on Dataset C3d. The EER position is marked by a circle in the plot (EER = 0.31%).**

Table 6.16 FMR and FNMR Values Obtained for the Application of Recognition Techniques Designed for Touch-Based Recognition Systems to Touch-Equivalent Images Computed Using Proposed Three-Dimensional Method C and the Unwrapping Technique on Dataset C3d

EER (%)	FNMR (%) at FMR = 0.01%	FNMR (%) at FMR = 0.10%
0.31	0.91	0.40

6.3.3.3 Comparison of Different Technologies

The best results obtained using the proposed touchless approaches on Dataset C3d were compared with those obtained using the NIST BOZORTH3 algorithm [186] on the touch-based images of Dataset Ct. The methods that were compared with the evaluated recognition system based on touchless acquisitions consisted of unwrapped three-dimensional samples obtained using three-dimensional Method C and the proposed multimodal technique based on two-dimensional samples.

Figure 6.18 presents the DET curves obtained using the analyzed biometric recognition systems. The results obtained using these methods at different points of the DET curve are summarized in Table 6.17. The results reported in Table 6.17 are ranked according to their EER values. The best EER was obtained using the proposed approach based on three-dimensional samples.

Among the compared methods, the proposed approach based on three-dimensional fingerprint samples achieved the greatest accuracy in the operative regions characterized by a low number of false acceptances. This result suggests that fingerprint recognition systems based on touchless multiple-view acquisitions could be effectively adopted in high-security applications. However, the evaluated touch-based system achieved better performance in other regions of the DET curve.

To better evaluate the applicability of the proposed approaches in a real application context, we estimated the confidence limits of the proposed method that obtained the most accurate results using two well-known techniques described in the literature. The first technique assumed that the obtained data were normally distributed, whereas the second technique was based on a bootstrap approach. These techniques are described in Section 2.3.4.

The performed tests were conducted on the results of the proposed approach based on touch-equivalent images obtained using three-dimensional Method C with Dataset C3d. All results were within a confidence level equal to 90%. As suggested in [65], the bootstrap technique was applied for 1000 iterations.

Figure 6.19 presents the obtained DET curves, and Table 6.18 presents the estimated confidence boundaries of the EER point.

Figure 6.19 and Table 6.18 indicate that the figures of merit computed for evaluating the biometric approach described the system accuracy with small confidence boundaries. Thus, the proposed approach should also yield satisfactory results for larger biometric datasets.

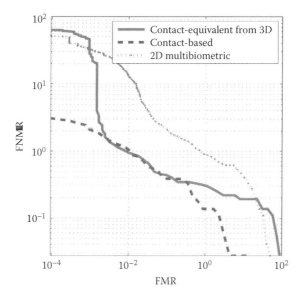

Figure 6.18 DET curves obtained using different recognition technologies in the scenario evaluation: the proposed approach based on unwrapping three-dimensional models, a touch-based recognition system, and the proposed multimodal system based on two-dimensional samples.

6.3.4 Speed

The computational time required for the studied approaches for touchless fingerprint recognition based on two-dimensional and three-dimensional samples and touch-based biometric systems were compared.

All presented methods were written in Matlab language (R2011b 64 bit) on an Intel Xeon 3.30 GHz running Windows 7 Professional 64 bit. The considered

Table 6.17 FMR and FNMR Values Obtained Using Different Recognition Technologies in the Scenario Evaluation: The Proposed Approach Based on the Unwrapping of Three-Dimensional Models, a Touch-Based Recognition System, and the Proposed Multimodal System Based on Two-Dimensional Samples

Method	EER (%)	FNMR (%) at FMR = 0.01%	FNMR (%) at FMR = 0.10%
Contact equivalent from 3D	0.31	0.91	0.40
Contact based	0.32	1.03	0.38
2D multimodal	0.90	12.71	2.06
2D single view	1.32	5.52	2.99

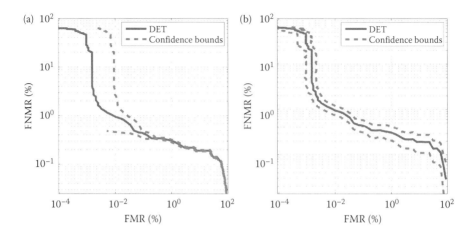

Figure 6.19 Confidence limits of the DET curve obtained using the proposed approach based on unwrapping three-dimensional models with Dataset C3d: (a) the confidence estimated assuming a normal distribution and (b) the confidence estimated using the bootstrap technique.

Table 6.18 Confidence Limits of the EER Obtained Using the Proposed Approach Based on Unwrapping Three-Dimensional Models with Dataset C3d

Confidence Estimation Algorithm	EER (%)	min FMR (%)	max FMR (%)	min FNMR (%)	max FNMR (%)
Normal distribution	0.31	0.21	0.41	0.30	0.32
Bootstrap	0.31	0.30	0.32	0.20	0.42

implementations were not optimized in terms of computational complexity and did not use parallel computing strategies.

The evaluated touchless techniques first computed touch-equivalent fingerprint images and then applied the same feature extraction and matching algorithms used for touch-equivalent images. Compared with biometric recognitions performed using touch-based systems, the recognitions based on touchless samples required an additional time t_e for computing touch-equivalent images. In the case of systems based on two-dimensional samples, t_e was approximately 2.70 s. Considering systems based on three-dimensional data, the time interval t_e can be divided into the time t_{3D} required to perform the three-dimensional reconstruction of the finger surface and the time t_u required for the unwrapping step. The values of t_{3D} and t_u were approximately 28.47 and 8.25 s, respectively. Approximately 27.35% of the time required for the proposed three-dimensional reconstruction technique is for the estimation of the corresponding pairs of points in the images related to the different views of the acquisition setup.

Moreover, the proposed techniques are designed to be easily parallelizable. For example, a parallel implementation of the proposed methods based on CUDA techniques [337] would drastically decrease the required computational time. A parallel implementation would most likely allow the proposed techniques based on three-dimensional samples to be used in real-time biometric applications.

6.3.5 Cost

The costs of touch-based biometric sensors for fingerprint recognition systems vary. For example, swipe sensors integrated in mobile devices or personal computers are approximately $10, while optical area scan sensors are more than $1000. The price of the sensor is determined by the acquisition technology, accuracy, and number of fingerprints that can be simultaneously captured.

The touch-based sensor used to perform the proposed scenario evaluation was a CrossMatch Verifier V300 [333], which is an area scan sensor based on optical technologies that costs approximately $700.

The proposed touchless acquisition systems are based on different hardware setups with different costs. These systems use one or two cameras and different illumination techniques. We used SX90CR CCD cameras with 25-mm Tamron lenses, which cost approximately $1500. The price of the LED illuminators used was less than $100.

The final cost of the proposed hardware setups was higher than that of many touch-based fingerprint sensors. However, the reported prices are for prototypal hardware configurations. Commercial versions of the proposed touchless recognition systems would employ less-expensive cameras.

Similar to touch-based recognition systems, touchless sensors can also use low-cost hardware configurations. For example, the system described in [262] is based on a single webcam. Cameras integrated in mobile devices [264] can also be used, eliminating the hardware costs of cameras.

6.3.6 Scalability

Touch-based fingerprint recognition systems are characterized by high scalability. In fact, the largest biometric datasets described in the literature are composed of fingerprint samples. Moreover, the existing AFIS can recognize millions of templates [29].

An important goal of studies of touchless fingerprint recognition systems is to guarantee interoperability with the existing AFIS to permit an AFIS system to work effectively with fingerprints acquired and processed using different sensors and techniques (e.g., live and latent fingerprints).

The proposed touchless fingerprint techniques that compute touch-equivalent images can use many modules of traditional fingerprint recognition systems: feature extraction techniques, matching methods, databases, and network infrastructures.

Moreover, these techniques can perform acquisitions based on webcams or on cameras integrated in mobile devices, increasing the possible diffusion of fingerprint recognition systems. Therefore, touchless techniques can improve the scalability of fingerprint recognition systems.

6.3.7 Interoperability

An important goal of studies of touchless fingerprint recognition systems is to guarantee interoperability with the existing AFIS. This property allows an AFIS system to effectively work with fingerprints acquired and processed using different sensor and techniques (e.g., live and latent fingerprints).

To evaluate the compatibility of touch-equivalent fingerprint images obtained using the proposed approach based on three-dimensional models with the existing AFIS, we evaluated the performance of a well-known recognition technique using a dataset comprising both touch-equivalent images and touch-based images. The used dataset was composed of touch-equivalent images obtained from Dataset C3d and touch-based images pertaining to Dataset Ct. The adopted matching algorithm was NIST BOZORTH3 [186].

Figure 6.20 presents the achieved DET curve. The reported results are based on 4,324,320 identity comparisons. The obtained EER was 1.65%. The performed test achieved less accurate results compared to those achieved separately using the matching method on the single datasets.

Considering that touch-equivalent images obtained from three-dimensional models do not exhibit nonlinear distortions due to different finger pressures on the sensor platen and that these images can present artifacts introduced by the three-dimensional reconstruction and unwrapping process, genuine identity comparisons performed between touch-based and touch-equivalent samples can obtain reduced performance due to a lower number of matched minutiae. We therefore evaluated the matching scores for genuine samples obtained using the considered acquisition techniques. Figure 6.21 presents the functions describing the matching scores for genuine samples acquired using touchless and touch-based sensors, and Table 6.19 reports the mean matching scores.

The matching scores for genuine samples acquired using different sensors were approximately half those achieved for samples acquired using the same technique. Therefore, touch-equivalent images and touch-based fingerprint images are partially compatible. To increase the interoperability between touchless and touch-based technologies, future studies should model the distortions present in the considered types of fingerprint images.

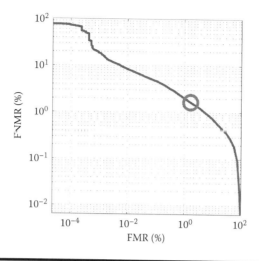

Figure 6.20 DET curve obtained using the method NIST BOZORTH3 [186] on a dataset composed of both touch-equivalent and touch-based images. The EER position is marked by a circle in the plot (EER = 1.65%). (Adapted from C. I. Watson et al., *User's Guide to NIST Biometric Image Software (NBIS)*, 2007.)

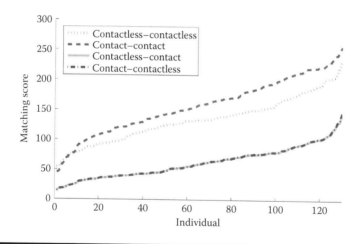

Figure 6.21 Matching scores for genuine samples captured using touchless and touch-based techniques.

6.3.8 Usability

Similar to the work reported in [44], we performed a preliminary usability evaluation according to ISO 9241-11 [338]. Usability is defined as "the extent to which a product can be used by specified users to achieve specified goals with effectiveness,

Table 6.19   Mean matching Scores for Genuine Samples Captured Using Touchless and Touch-Based Techniques

Comparison	Mean Matching Score between Genuines
Touchless–touchless	133.625
Touch–touch	157.163
Touch–touchless	62.288
Touchless–touch	62.349

efficiency, and satisfaction in a specified context of use." This standard considers three areas of measurement: efficiency, effectiveness, and user satisfaction.

1. Efficiency quantifies the resources expended to achieve accuracy and completeness. Efficiency is usually measured in terms of time.
2. Effectiveness measures the accuracy and completeness with which users achieve specified goals. Important evaluated aspects are the completion rate and the number of errors.
3. User satisfaction is subjective and is the degree to which the considered technology meets the users' expectations. The evaluation of user satisfaction should consider the ease of use and the usefulness of the technology.

The 13 volunteers evaluated usability during the creation of Dataset C3d and Dataset Ct based on a scenario evaluation regarding access control in a laboratory. This test aimed at comparing the usability of three-dimensional Method C and a biometric system with a touch-based acquisition sensor CrossMatch V300 [333].

Dataset C3d was created without using techniques that automatically captured sufficient-quality frames during the live acquisitions, and the best-quality frames representing the captured fingerprints were selected by a skilled operator during each acquisition because our implementation of the proposed technique for searching for the best-quality frames in frame sequences describing touchless fingerprint acquisitions (Section 5.2) does not function in real-time because this software is a prototype developed in Matlab [339]. To obtain comparable data, the images pertaining to Dataset C3t were also captured without using automatic quality-evaluation methods. However, commercial implementation of the proposed quality-assessment technique for touchless fingerprint images should permit the adoption of a quality-assessment method using real-time acquisition systems.

Considering the variability introduced by the human operator in every biometric acquisition, we proposed a preliminary qualitative evaluation of efficiency aspects focused on quantitative measurements of the time requirement for every biometric acquisition. As in [44,340], the first considered aspect was the time required for proper placement of the finger on the acquisition sensor, which is extremely similar to that of the evaluated touchless and touch-based fingerprint recognition technologies. The training time was also evaluated by measuring the time required to teach the users in a verbal manner. Touchless acquisition techniques required that the finger placement be described to the users to prevent out-of-focus problems. Touch-based acquisitions required a description of the proper pressure that should be applied to the sensor platen. The results indicated similar training times for the touchless and touch-based sensors. However, the training for the considered touch-based fingerprint sensor was not necessary for five volunteers because these volunteers had previously attempted traditional fingerprint recognition systems. These cases were not considered in the evaluation of the mean training time required for the touch-based fingerprint recognition system.

Similar to [44], the effectiveness evaluation was performed by analyzing the quality of the captured images. To obtain comparable results, we applied the quality evaluation method NIST NFIQ [221] to the touch-based samples pertaining to Dataset Ct, to the touch-equivalent images obtained by applying three-dimensional Method C on Dataset C3d, and to the touch-equivalent images obtained using the proposed approach for processing single touchless images on Dataset C3d-1 and on Dataset C3d-2. This method returns five quality levels: (5) poor, (4) fair, (3) good, (2) very good, and (1) excellent. Table 6.20 reports the number of images in the evaluated datasets with quality levels equal to or greater than 2. In the performed experiment, touchless acquisition techniques produced fewer low-quality images than touch-based acquisition systems. All of the touch-equivalent images obtained from Dataset C3d were characterized as good and very good quality. Moreover, the number of touch-equivalent images pertaining to Dataset C3d-1 and Dataset C3d-2 with quality equal to (2) very good or (1) excellent was higher than that obtained for touch-based Dataset Ct.

To evaluate user satisfaction, each volunteer was asked to perform a satisfaction survey after completing the biometric acquisitions. The following questions were included in the form:

- Q1—Is the acquisition procedure comfortable?
- Q2—What do you think about the time needed for every acquisition?

Table 6.20 Effectiveness Comparison of Touchless and Touch-Based Systems Based on NIST NFIQ Software

Dataset	Percentage of Images with Quality ≥2
Dataset C3d	100.000
Dataset C3t	95.000
Dataset C3d-1	99.423
Dataset C3d-2	99.231

The following responses were possible: (1) very poor, (2) poor, (3) sufficient, (4) good, and (5) excellent. The questions were related to the used systems and two new envisioned technologies:

1. The tested touch-based acquisition sensor
2. The tested two-view acquisition technique of three-dimensional Method C
3. A proposed touchless acquisition system similar to the hardware setup used by three-dimensional Method C but with inverted finger placement (rotated by 180°)
4. The proposed touchless acquisition system based on active cameras

The envisioned acquisition system based on active cameras consisted of a calibrated multiple-view setup that could be moved in three-dimensional space by stepper motors. The cameras were fixed on a support that was dynamically moved by the motors according to the finger position. The movements were performed by evaluating the camera focus in the captured images.

Figure 6.22 presents the responses to the questionnaire for the different acquisition technologies.

The mean values of the responses to question Q1 were as follows: (I) 4.00, (II) 3.83, (III) 4.50, and (IV) 4.63. The best results were obtained for the envisioned system based on three-dimensional Method C with inverse finger placement. Then, a two-sample t-test assuming unequal variances was conducted to compare the responses for the envisioned system based on three-dimensional Method C with inverse finger placement with those for the implemented touch-based acquisition systems. In this case, the null hypothesis assumed that the two samples used to process the mean values belonged to the same distribution; thus, no difference in user opinion regarding the two tested systems was observed. This result was significant at the 0.044 level and beyond, indicating that the null hypothesis could be rejected with confidence.

The mean values of the responses to question Q2 were as follows: (I) 3.92, (II) 3.92, (III) 4.25, and (IV) 4.00. In addition, a two-sample t-test assuming unequal variances was conducted to compare the responses for the envisioned system based on three-dimensional Method C with inverse finger placement with those for the implemented touch-based acquisition systems. A difference in the scores for the two conditions was observed; this difference was equal to $P = 0.336$. This result suggests that collecting more samples to properly analyze this aspect should be useful.

The usability evaluation thus obtained satisfactory results. The efficiency of the proposed touchless approach is comparable to that of touch-based biometric systems. The proposed touchless recognition technique based on three-dimensional samples was more effective than the evaluated touch-based biometric system. Moreover, the use of active cameras may increase user satisfaction compared with touch-based systems.

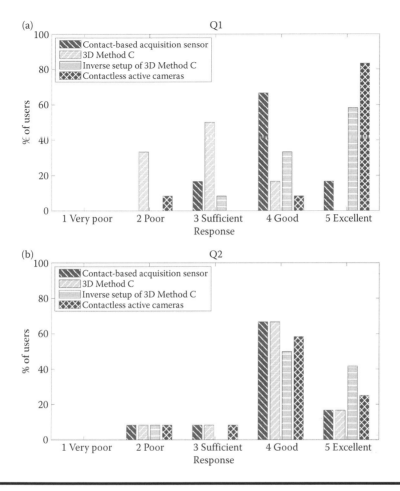

Figure 6.22 Usability comparison of different technologies. A set of volunteers responded to the questions: (a) (Q1) "Is the acquisition procedure comfortable?"; (b) (Q2) "What do you think about the time needed for every acquisition?"

6.3.9 Social Acceptance

Similar to other studies in the literature [45], we compared the social acceptability of different biometric techniques by analyzing the answers to specific sets of questions.

After the acquisition of Dataset C3d and Dataset Ct, the volunteers were asked to respond to a first set of questions focused on the analysis of feelings about different aspects of touchless and touch-based fingerprint recognition systems and to a second set of questions that aimed to evaluate the final opinion of the users about touchless technologies.

The first set of questions included the following:

- Q3—Are you worried about hygiene issues?
- Q4—Are you worried about possible security lapses due to latent fingerprints?
- Q5—Do you think that biometric data could be improperly used for police investigations?
- Q6—Do you feel that the system invaded your privacy?

The possible responses were the following: (1) very worried, (2) worried, (3) normal, (4) not worried, and (5) high trust.

The mean of the obtained votes was computed for each question. The obtained results are presented in Table 6.21. The volunteers perceived touchless techniques as more hygienic than touch-based methods. Moreover, the volunteers were less worried regarding possible security lapses due to the release of latent fingerprints in the case of touchless acquisitions. In contrast, the evaluated techniques yielded similar results for the questions regarding privacy invasiveness and worries regarding possible improper uses of biometric data for police investigations.

Then, we conducted a two-sample t-test assuming unequal variances to compare the results obtained for question Q3 for touchless and touch-based systems. In this case, the null hypothesis assumed that the two samples used to process the mean values belonged to the same distribution; thus, no difference in user opinion regarding the two tested systems was observed. This result was significant at the $2.142E - 07$ level and beyond, indicating that the null hypothesis could be rejected with confidence.

Similarly, we conducted a two-sample t-test assuming unequal variances to compare the results for question Q4 for touchless and touch-based systems. In this case, the null hypothesis assumed that the two samples used to process the mean values belonged to the same distribution; thus, no difference in user opinion regarding the two tested systems was observed. This result was significant at the $6.115E - 06$ level and beyond, indicating that the null hypothesis could be rejected with confidence.

Figure 6.23 presents the histograms of the obtained results.

Table 6.21 Comparison of Social Acceptance Aspects of Different Technologies

		Mean Vote	
Question		Touch-Based	Touchless
Q3	Are you worried about hygiene issues?	2.500	4.667
Q4	Are you worried about possible security lapses due to latent fingerprints?	2.500	4.417
Q5	Do you think that biometric data could be improperly used for police investigations?	2.917	2.917
Q6	Do you feel the system invaded your privacy?	3.250	3.417

Note: The possible responses were (1) very worried, (2) worried, (3) normal, (4) not worried, and (5) high trust.

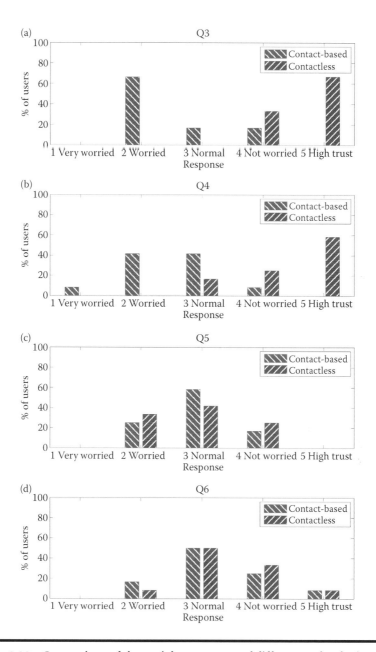

Figure 6.23 Comparison of the social acceptance of different technologies. A set of volunteers responded to the questions: (Q3) "Are you worried about hygiene issues?", (Q4) "Are you worried about possible security lapses due to latent fingerprints?", (Q5) "Do you think that biometric data could be improperly used for police investigations?", and (Q6) "Do you feel that the system invaded your privacy?"

The second set of questions was as follows:

- Q7 Would you be willing to use the touchless fingerprint biometric system daily?
- Q8 Do you prefer touchless systems to touch-based systems?

All volunteers responded favorably to the daily use of touchless fingerprint biometric system and preferred touchless systems.

These results suggest that touchless fingerprint recognition systems can achieve greater social acceptance than touch-based methods.

6.3.10 Security

Compared with touch-based fingerprint recognition systems, touchless techniques do not present security lapses due to the release of latent fingerprints during the acquisition step.

Although the proposed biometric recognition methods do not include a vitality detection module, techniques based on infrared illumination methods described in the literature could be integrated into these methods [306].

6.3.11 Privacy

The level of privacy compliance of touch-equivalent recognition techniques is extremely similar to that of touch-based biometric systems. In fact, specific strategies for privacy protection must be adopted based on the application scenario. Moreover, privacy protection methods described in the literature for touch-based fingerprint recognition systems could be applied to the proposed touchless fingerprint recognition techniques to ensure privacy-compliant systems.

6.3.12 Final Results

The results obtained using the proposed touchless fingerprint recognition techniques in the performed scenario evaluation can be summarized as follows:

1. *Accuracy*: The proposed method based on two-dimensional samples achieved satisfactory results and should be effective in different application contexts. In high-security applications, the proposed approach based on three-dimensional samples can achieve better accuracy than touch-based fingerprint recognition systems.
2. *Speed*: The proposed touchless recognition techniques require additional computational time compared to touch-based methods. In particular, the performance of the three-dimensional reconstruction technique could be improved using parallel computing strategies. A commercial implementation of the proposed approaches would likely be usable in real-time applications.

3. *Cost*: The proposed touchless acquisition techniques are prototypes based on industrial cameras, which should be substituted with less-expensive devices to reduce costs compared to touch-based acquisition sensors.

4. *Scalability*: Touchless fingerprint recognition systems can provide greater scalability compared with touch-based methods because these systems employ acquisition hardware previously adopted for different applications (e.g., cameras integrated in mobile devices and webcams).

5. *Interoperability*: The proposed touchless fingerprint recognition methods are partially compatible with the existing AFIS. Studies of skin deformation due to contact with the sensor platen should improve the interoperability between touchless and touch-based techniques.

6. *Usability*: The proposed touchless acquisition technique for computing three-dimensional samples is more effective than the evaluated touch-based method. The efficiency of the two techniques is comparable. In contrast, the user satisfaction associated with the proposed acquisition setup is lower than that achieved using touch-based sensors. However, the envisioned systems based on three-dimensional Method C with an inverted finger orientation and the envisioned systems based on active cameras yielded better results for user satisfaction compared with touch-based techniques.

7. *Social acceptance*: Using touchless acquisition techniques can increase the user acceptance of fingerprint recognition systems. In particular, users appreciate the absence of latent fingerprints and hygienic improvements.

8. *Security*: Touchless acquisition can increase the security of fingerprint recognition systems because no latent fingerprint are released on the sensor surface.

9. *Privacy*: The privacy compliance of touchless and touch-based fingerprint recognition systems is comparable. Privacy protection techniques should be applied in all systems depending on the application context.

6.4 Computation of Synthetic Three-Dimensional Models

The proposed method for the computation of synthetic three-dimensional fingerprint models (Section 5.4) was evaluated by computing a set of simulated fingerprint models from rolled fingerprint images and by comparing the resulting models with the corresponding touchless images captured using the acquisition setup of three-dimensional Method B (Section 5.3.2). The reference touchless images were captured using Sony XCD-SX90CR cameras and had a size of 1280×960 pixels. Rolled fingerprints were chosen over live touch-based images because rolled samples describe a wider area of the fingertip.

The following parameters were used to simulate three-dimensional models of finger shape: $a_W = 3/5$, $a_{H1} = 4$, and $a_{H2} = 5$. The parameter used to normalize

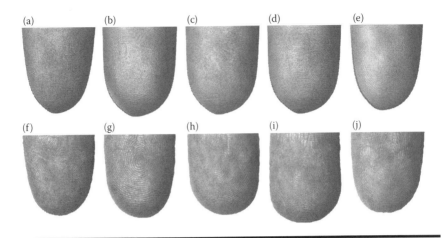

Figure 6.24 Examples of simulated fingerprint samples: (a–e) simulated fingerprint models shown from a top view and (f–j) touchless real images. Each simulated fingerprint model was computed starting from a touch-based image I_C acquired from the same finger used to perform real touchless acquisitions. The camera was positioned identically to the simulated point of views used to process the virtual images. (R. Donida Labati et al., Accurate 3D fingerprint virtual environment for biometric technology evaluations and experiment design, in *Proceedings of the IEEE International Conference on Computational Intelligence and Virtual Environments for Measurement Systems and Applications*, pp. 43–48. © 2013, IEEE.)

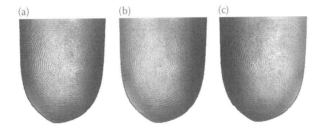

Figure 6.25 Examples of synthetic models computed using different positions of a punctiform light source: (a) light positioned to the left, (b) light in a central position, and (c) light positioned to the right. The results indicate that the studied method can be used to rapidly and effectively evaluate experimental design by simulating different conditions and geometries of the setup (cameras, illumination systems, and fingers). (R. Donida Labati et al., Accurate 3D fingerprint virtual environment for biometric technology evaluations and experiment design, in *Proceedings of the IEEE International Conference on Computational Intelligence and Virtual Environments for Measurement Systems and Applications*, pp. 43–48. © 2013, IEEE.)

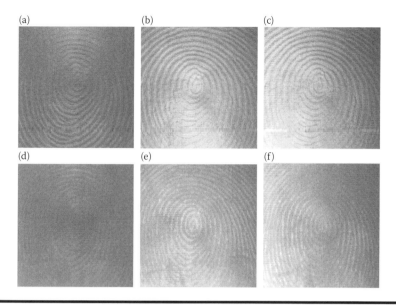

Figure 6.26 Examples of portions of touchless three-dimensional models obtained by the implemented approach for simulating different illumination conditions: (a–c) synthetic models obtained using our approach and (d–f) the corresponding touchless images. The images (a, d) are based on illumination from the left side, (b, e) from the right side, and (c, f) from the top side. (R. Donida Labati et al., Virtual environment for 3-D synthetic fingerprints, in *Proceedings of the IEEE International Conference on Virtual Environments, Human-Computer Interfaces and Measurement Systems*, pp. 48–53. © 2012, IEEE.)

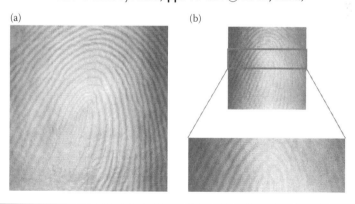

Figure 6.27 Examples of portions of touchless three-dimensional models obtained by simulating a fingerprint captured using: (a) a Sony XCD-SX90CR camera and (b) a VGA webcam. (R. Donida Labati et al., Virtual environment for 3-D synthetic fingerprints, in *Proceedings of the IEEE International Conference on Virtual Environments, Human-Computer Interfaces and Measurement Systems*, pp. 48–53. © 2012, IEEE.)

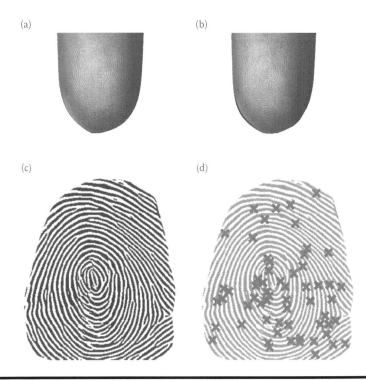

Figure 6.28 Simulation of a two-view acquisition, computation of the three-dimensional model, unwrapping, and computation of touch-equivalent images: (a) simulated acquisition obtained by camera A and (b) by camera B; (c) a touch-equivalent image computed from the synthetic model after reconstructing the three-dimensional model from the simulated two-view acquisition and the unwrapping step; and (d) extraction of the minutiae from the touch-equivalent image. These results indicate that the synthetic samples obtained using the proposed method can be effectively used to evaluate biometric recognition algorithms. (R. Donida Labati et al., Accurate 3D fingerprint virtual environment for biometric technology evaluations and experiment design, in *Proceedings of the IEEE International Conference on Computational Intelligence and Virtual Environments for Measurement Systems and Applications*, pp. 43–48. © 2013, IEEE.)

the intensities of the ridges and valleys was $\Delta_n = 0.2$. The sizes and standard deviations of the Gaussian filters that were used to perform the focus simulation were as follows: $m = \{10, 7, 5, 3, 2\}$ and $\sigma = \{7, 5, 3, 2, 1\}$.

To evaluate the realism of the proposed illumination model, we simulated acquisitions performed using a white LED light to enhance the visibility of the ridge pattern. Both an ambient light and an LED light source were simulated. Examples of the results are presented in Figure 6.24, which illustrates that the computed models are realistic and resemble real touchless acquisitions.

Then, different positions of an LED light were simulated (Figures 6.25 and 6.26). The results obtained indicated that the proposed method can realistically simulate different illumination conditions.

We also evaluated the results obtained by simulating acquisitions obtained using different types of cameras. As an example, Figure 6.27 presents portions of images obtained by simulating frontal acquisitions acquired using a Sony XCD-SX90CR camera (Figure 6.27a) and by a VGA webcam (Figure 6.27b). The image depicted in Figure 6.27b was obtained by using different parameters of the proposed method, which permitted the simulation of more blur, out-of-focus regions, and noise. This image effectively resembles acquisitions performed by low-cost cameras.

To analyze the capability of the proposed simulation method to create sets of touchless fingerprint images usable for testing biometric recognition algorithms, we analyzed the results achieved using the proposed methods for three-dimensional reconstruction and processing on simulated acquisitions. First, the proposed method was applied to simulate two-view images captured using the acquisition setup of three-dimensional Method B (Section 5.3.2). Then, the three-dimensional reconstruction procedure and unwrapping method described in Section 5.3.2 were applied, and the minutiae were extracted using the software NIST MINDTC [186]. Examples of the results obtained are presented in Figure 6.28 and are very similar to those obtained for real samples. Thus, the presented simulation method could be used to create reference data for testing novel feature extraction and matching algorithms.

6.5 Summary

First, we evaluated the accuracy of the proposed techniques for touchless systems based on two-dimensional samples. The experiments were performed using datasets of touchless fingerprint images captured in our laboratory using different hardware setups. The evaluated techniques were the proposed methods for assessing the quality of touchless fingerprint images and the presented approach for estimating the core point in touchless images. The methods for assessing the quality of touchless images indicated good performance, and these experiments demonstrated that these techniques could be effectively used to identify the best-quality frames in frame sequences describing a finger moving toward the camera. In addition, the core-detection approach achieved satisfactory results for touchless and touch-based fingerprint images. The evaluation of the method for compensating for perspective distortions and finger rotations also obtained satisfactory results, indicating significant increased accuracy compared to the baseline approach.

Then, the proposed approaches based on three-dimensional fingerprint samples were analyzed. The results obtained by the different three-dimensional reconstruction strategies were encouraging for all of the proposed techniques. Next, the quality of the touch-based images obtained by unwrapping three-dimensional models was analyzed, which revealed that the proposed method for assessing the quality of touch-equivalent images yielded results with increased accuracy compared with methods described in the literature for touch-based techniques.

Next, recognition techniques based on touchless samples and traditional touch-based systems were compared. This comparison included the following aspects of biometric technologies: (I) accuracy, (II) speed, (III) cost, (IV) scalability, (V) interoperability, (VI) usability, (VII) social acceptance, (VIII) security, and (IX) privacy.

The performed accuracy evaluation indicated that, for high-accuracy applications, the proposed method based on unwrapping three-dimensional models is more accurate than touch-based recognition techniques. Systems based on touchless two-dimensional samples also obtained satisfactory accuracy. The compatibility between touch-equivalent and touch-based fingerprint images was then analyzed. The results were encouraging but only demonstrated partial compatibility between touch-equivalent images and the existing AFIS. Moreover, the proposed approaches achieved improvements in scalability, usability, social acceptance, and security compared with traditional techniques.

Finally, the results obtained using the proposed technique for computing synthetic touchless samples were discussed. The performed tests indicated that this method simulated realistic data.

Chapter 7

Conclusions and Future Work

7.1 Conclusions

This book presented innovative multidisciplinary technologies for touchless fingerprint recognition compatible with less-constrained scenarios compared with traditional biometric systems. Their design and realization required a multidisciplinary approach involving optical acquisition systems, multiple-view geometry, image processing, pattern recognition, computational intelligence, statistics, and cryptography.

The studied technologies encompass all aspects of touchless biometric recognition systems based on two-dimensional and three-dimensional fingerprint samples, including methods for acquiring and processing biometric data. All modules of biometric systems were studied and implemented: hardware setups, acquisition techniques, biometric sample quality assessment, reduction of the effects of perspective distortions and finger rotations, three-dimensional reconstruction of fingerprint models, computation of touch-equivalent images, and feature extraction and matching. A method for computing synthetic touchless fingerprint samples was also developed to accelerate system testing.

The implemented methods were evaluated using datasets composed of touchless biometric samples captured in different application contexts. Satisfactory results were obtained that demonstrated the applicability of touchless fingerprint recognition technologies in different application scenarios.

The realized touchless biometric technologies were also compared with traditional touch-based fingerprint recognition systems in a scenario evaluation of the following aspects of biometric technologies: accuracy, speed, cost, scalability, interoperability, usability, social acceptance, security, and privacy. The results indicated that in high-security contexts, touchless biometric systems based on three-dimensional samples could achieve greater accuracy than traditional touch-based fingerprint recognition systems. For example, the realized system based on three-dimensional samples achieved an EER of 0.31% for a dataset of 1040 samples, whereas the evaluated touch-based system achieved an EER of 0.32% for a dataset composed of the same number of samples captured from the same individuals. Satisfactory results were also achieved using the implemented technique based on touchless two-dimensional samples, which obtained sufficient accuracy for use in many low-cost applications (e.g., in mobile devices), with an EER of 1.32%. During the scenario evaluation, the compatibility of touch-equivalent images obtained from three-dimensional models with existing biometric databases was also analyzed, and encouraging results were obtained. Furthermore, the researched approaches featured improved scalability, usability, social acceptance, and security compared with traditional touch-based techniques.

Although the analysis of the performance of the implemented recognition technologies based on three-dimensional features yielded valuable results, further improvements could be obtained in future studies focusing on template alignment strategies.

Therefore, the evaluations demonstrated that touchless fingerprint recognition methods can be used effectively and advantageously in different application scenarios in which live touch-based techniques are employed.

7.2 Future Work

One of the goals of the technologies described in this book is to increase the usability and acceptability of biometric systems. The usability of the realized approaches might be further improved by acquisition setups based on active cameras to perform acquisitions in a more comfortable manner.

The compatibility of the implemented techniques with the existing AFIS could also be improved to permit the diffusion of these techniques in government and investigative applications, such as by creating touch-equivalent images that can better simulate the skin distortion caused by the pressure of the finger on the sensor.

To increase the accuracy of fingerprint recognition systems, further studies of feature extraction and matching methods based on three-dimensional feature sets might also be performed to exploit additional information related to finger volume and create new distinctive features in three-dimensional space. To increase the accuracy of these techniques, three-dimensional reconstruction techniques that are more robust to noisy data may also be investigated.

To demonstrate the applicability of the proposed approaches in different contexts and environmental conditions, new datasets may be collected for testing.

Commercial implementations of the presented approaches should also drastically reduce the computational time required by the implemented techniques. For example, parallel-processing architectures based on graphic processing units (GPU) might permit real-time applications.

References

1. International Organization for Standards, ISO/IEC JTC1 SC37 Standing Document 2, version 8, Harmonized Biometric Vocabulary, August 1997.
2. A. Jain, A. Ross, and S. Prabhakar, An introduction to biometric recognition, *IEEE Transactions on Circuits and Systems for Video Technology*, vol. 14, no. 1, pp. 4–20, 2004.
3. M. Meytlis and L. Sirovich, On the dimensionality of face space, *IEEE Transactions on Pattern Analysis and Machine Intelligence*, vol. 29, no. 7, pp. 1262–1267, 2007.
4. D. Muramatsu, M. Kondo, M. Sasaki, S. Tachibana, and T. Matsumoto, A Markov chain Monte Carlo algorithm for Bayesian dynamic signature verification, *IEEE Transactions on Information Forensics and Security*, vol. 1, no. 1, pp. 22–34, 2006.
5. D. Maltoni, D. Maio, A. K. Jain, and S. Prabhakar, *Handbook of Fingerprint Recognition*, 2nd Edition. Springer Publishing Company, London, UK, 2009.
6. R. Donida Labati and F. Scotti, Fingerprint, in H. van Tilborg and S. Jajodia, Eds., *Encyclopedia of Cryptography and Security*, 2nd edition. Springer, New York, 2011, pp. 460–465.
7. M. J. Burge and K. W. Bowyer, A. S. Li, Ed., *Handbook of Iris Recognition*. Springer Publishing Company, London, UK, 2013.
8. J. Daugman, How iris recognition works, in *Proceedings of the International Conference on Image Processing*, vol. 1, 2002, pp. 33–36.
9. R. Gross, Face databases, in A. S. Li, Ed., *Handbook of Face Recognition*, Springer, New York, 2005.
10. N. Duta, A survey of biometric technology based on hand shape, *Pattern Recognition*, vol. 42, no. 11, pp. 2797–2806, 2009.
11. A. Kong, D. Zhang, and M. Kamel, A survey of palmprint recognition, *Pattern Recognition*, vol. 42, pp. 1408–1418, 2009.
12. D. Zhang, Z. Guo, G. Lu, L. Zhang, Y. Liu, and W. Zuo, Online joint palmprint and palmvein verification, *Expert Systems with Applications*, vol. 38, no. 3, pp. 2621–2631, 2011.
13. A. Pflug and C. Busch, Ear biometrics: A survey of detection, feature extraction and recognition methods, *IET Biometrics*, vol. 1, no. 15, pp. 114–129, June 2012.
14. B. Bhanu and H. Chen, *Human Ear Recognition by Computer*, 1st Edition. Springer Publishing Company, London, UK, 2008.

15. H. Beigi, *Fundamentals of Speaker Recognition.* Springer Science+Business Media, New York, 2011.

16. J. Benesty, M. M. Sondhi, and Y. A. Huang, *Springer Handbook of Speech Processing.* Springer-Verlag, Secaucus, NJ, 2007.

17. D. Impedovo and G. Pirlo, Automatic signature verification: The state of the art, *IEEE Transactions on Systems, Man, and Cybernetics, Part C: Applications and Reviews*, vol. 38, no. 5, pp. 609–635, 2008.

18. J. Wang, M. She, S. Nahavandi, and A. Kouzani, A review of vision-based gait recognition methods for human identification, in *Proceedings of the International Conference on Digital Image Computing: Techniques and Applications*, 2010, pp. 320–327.

19. D. Shanmugapriya and G. Padmavathi, A survey of biometric keystroke dynamics: Approaches, security and challenges, *International Journal of Computer Science and Information Security*, vol. 5, pp. 115–119, 2009.

20. A. K. Jain, S. C. Dass, K. Nandakumar, and K. N, Soft biometric traits for personal recognition systems, in *Proceedings of the International Conference on Biometric Authentication*, 2004, pp. 731–738.

21. S. Denman, A. Bialkowski, C. Fookes, and S. Sridharan, Determining operational measures from multi-camera surveillance systems using soft biometrics, in *Proceedings of the IEEE International Conference on Advanced Video and Signal-Based Surveillance*, 2011, pp. 462–467.

22. R. Donida Labati, A. Genovese, V. Piuri, and F. Scotti, Weight estimation from frame sequences using computational intelligence techniques, in *Proceedings of the IEEE International Conference on Computational Intelligence for Measurement Systems and Applications*, July 2012, pp. 29–34.

23. S. Denman, C. Fookes, A. Bialkowski, and S. Sridharan, Soft-biometrics: Unconstrained authentication in a surveillance environment, in *Proceedings of Digital Image Computing: Techniques and Applications*, December 2009, pp. 196–203.

24. M. Demirkus, K. Garg, and S. Guler, Automated person categorization for video surveillance using soft biometrics, in *Proceedings of SPIE,* vol. 7667, 2010, pp. 76670P-76670P-12, 2010.

25. J. Tapia and C. Perez, Gender classification based on fusion of different spatial scale features selected by mutual information from histogram of LBP, intensity, and shape, *IEEE Transactions on Information Forensics and Security*, vol. 8, no. 3, pp. 488–499, 2013.

26. A. Dantcheva, N. Erdogmus, and J.-L. Dugelay, On the reliability of eye color as a soft biometric trait, in *Proceedings of the IEEE Workshop on Applications of Computer Vision*, January 2011, pp. 227–231.

27. RNCOS Business Consultancy Services, Electronics security: Global biometric forecast to 2012, 2010.

28. International Biometric Group, Biometrics market and industry report, BMIR 2009–2014, 2009, http://www.ibgweb.com/.

29. P. Komarinski, *Automated Fingerprint Identification Systems (AFIS).* Elsevier Academic Press, New York, 2005.

30. W. Zhao, R. Chellappa, A. Rosenfeld, and P. J. Phillips, Face recognition: A literature survey, *ACM Computing Surveys*, vol. 35, pp. 399–458, 2003.

31. Y. Du, E. Arslanturk, Z. Zhou, and C. Belcher, Video-based non-cooperative iris image segmentation, *IEEE Transactions on Systems, Man, and Cybernetics, Part B: Cybernetics*, vol. 41, no. 1, pp. 64–74, 2011.

32. R. Donida Labati and F. Scotti, Noisy iris segmentation with boundary regularization and reflections removal, *Image Vision and Computing*, vol. 28, no. 2, pp. 270–277, 2010.

33. T. Hicks and R. Coquoz, Forensic DNA evidence, in *Encyclopedia of Biometrics,* Springer-Verlag, New York, 2009, pp. 573–579.

34. D. J. Hurley, B. Arbab-zavar, and M. S. Nixon, The ear as a biometric, in *Handbook of Biometrics,* Springer-Verlag, New York, 2007.

35. A. Ross, K. Nandakumar, and A. K. Jain, *Handbook of Multibiometrics (International Series on Biometrics).* Springer, Secaucus, NJ, 2006.

36. A. Azzini, S. Marrara, R. Sassi, and F. Scotti, A fuzzy approach to multimodal biometric continuous authentication, *Fuzzy Optimization and Decision Making,* vol. 7, no. 3, pp. 243–256, 2008.

37. S. Cimato, M. Gamassi, V. Piuri, D. Sana, R. Sassi, and F. Scotti, Personal identification and verification using multimodal biometric data, in *Proceedings of the IEEE International Conference on Computational Intelligence for Homeland Security and Personal Safety,* October 2006, pp. 41–45.

38. M. Gamassi, V. Piuri, D. Sana, O. Scotti, and F. Scotti, A multi-modal multi-paradigm agent-based approach to design scalable distributed biometric systems, in *Proceedings of the IEEE International Conference on Computational Intelligence for Homeland Security and Personal Safety,* April 2005, pp. 65–70.

39. M. Gamassi, V. Piuri, D. Sana, and F. Scotti, A high-level optimum design methodology for multimodal biometric systems, in *IEEE International Conference on Computational Intelligence for Homeland Security and Personal Safety,* July 2004, pp. 117–124.

40. National Science & Technology Council (NSTC) Subcommittee on Biometrics, Biometric testing and statistics, http://www.biometrics.gov/documents/biotestingandstats.pdf, 2006.

41. B. Dorizzi, R. Cappelli, M. Ferrara, D. Maio, D. Maltoni, N. Houmani, S. Garcia-Salicetti, and A. Mayoue, Fingerprint and on-line signature verification competitions at icb 2009, in M. Tistarelli and M. Nixon, Eds., *Advances in Biometrics,* vol. 5558, Springer, Berlin, Heidelberg, 2009, pp. 725–732.

42. T. Mansfield, G. Kelly, D. Chandler, and J. Kane, Biometric product testing final report V1.0, National Physical Laboratory, London, March 2001.

43. R. Ryan, The importance of biometric standards, *Biometric Technology Today,* vol. 2009, no. 7, pp. 7–10, 2009.

44. M. Theofanos, B. Stanton, C. Sheppard, R. Micheals, N. Zhang, W. Wydler, L. Nadel, and R. Rubin, Usability testing of height and angles of ten-print fingerprint capture, National Institute of Standards and Technology, June 2008.

45. M. El-Abed, R. Giot, B. Hemery, and C. Rosenberger, A study of users' acceptance and satisfaction of biometric systems, in *Proceedings of the IEEE International Carnahan Conference on Security Technology,* October 2010, pp. 170–178.

46. N. K. Ratha, J. H. Connell, and R. M. Bolle, An analysis of minutiae matching strength, in *Using Linear Symmetry Features as a Pre-processing Step for Fingerprint Images,* 2001, pp. 223–228.

47. F. Sabena, A. Dehghantanha, and A. Seddon, A review of vulnerabilities in identity management using biometrics, in *Second International Conference on Future Networks (CFN),* pp. 42–49, January 2010.

48. J. Galbally, J. Fierrez, J. Ortega-Garcia, and R. Cappelli, Fingerprint anti-spoofing in biometric systems, in S. Marcel, M. S. Nixon, and S. Z. Li, Eds., *Handbook of Biometric Anti-Spoofing,* Springer, London, 2014, pp. 35–64.

49. S. De Capitani di Vimercati, S. Foresti, and P. Samarati, Managing and accessing data in the cloud: Privacy risks and approaches, in *Proceedings of the 7th International Conference on Risks and Security of Internet and Systems,* October 2012.

50. S. Foresti, *Preserving Privacy in Data Outsourcing*. Springer, New York, 2011.

51. P. Samarati and S. De Capitani di Vimercati, Data protection in outsourcing scenarios: Issues and directions, in *Proceedings of the 5th ACM Symposium on Information, Computer and Communications Security*, April 2010.

52. E. Damiani, S. De Capitani di Vimercati, and P. Samarati, New paradigms for access control in open environments, in *Proceedings of the 5th IEEE International Symposium on Signal Processing and Information*, December 2005.

53. S. De Capitani di Vimercati, P. Samarati, and S. Jajodia, Policies, models, and languages for access control, in *Proceedings of the Workshop on Databases in Networked Information Systems*, March 2005.

54. C. Blundo, S. Cimato, S. De Capitani di Vimercati, A. De Santis, S. Foresti, S. Paraboschi, and P. Samarati, Efficient key management for enforcing access control in outsourced scenarios, in *Proceedings of the 24th IFIP TC-11 International Information Security Conference*, May 2009.

55. R. Donida Labati, V. Piuri, and F. Scotti, Biometric privacy protection: Guidelines and technologies, in M. S. Obaidat, J. Sevillano, and F. Joaquim, Eds., *Communications in Computer and Information Science*, vol. 314, Springer, New York, 2012, pp. 3–19.

56. S. Cimato, M. Gamassi, V. Piuri, R. Sassi, and F. Scotti, Privacy in biometrics, in *Biometrics: Theory, Methods, and Applications*, Wiley-IEEE Press, Haboken, New Jersey, 2008.

57. A. K. Jain, K. Nandakumar, and A. Nagar, Biometric template security, *EURASIP Journal on Advances in Signal Processing*, vol. 2008, pp. 1–17, 2008.

58. S. Cimato, R. Sassi, and F. Scotti, Biometric privacy, in H. van Tilborg and S. Jajodia, Eds., *Encyclopedia of Cryptography and Security*, 2nd Edition. Springer, New York, 2011, pp. 101–104.

59. S. Cimato, M. Gamassi, V. Piuri, R. Sassi, and F. Scotti, Privacy issues in biometric identification, *Touch Briefings*, pp. 40–42, 2006.

60. S. Cimato, R. Sassi, and F. Scotti, Biometrics and privacy, *Recent Patents on Computer Science*, vol. 1, pp. 98–109, June 2008.

61. A. K. Jain, R. M. Bolle, and S. Pankanti, *Biometrics: Personal Identification in Networked Society*. Springer, New York, 2005.

62. M. Gamassi, M. Lazzaroni, M. Misino, V. Piuri, D. Sana, and F. Scotti, Quality assessment of biometric systems: A comprehensive perspective based on accuracy and performance measurement, *IEEE Transactions on Instrumentation and Measurement*, vol. 54, no. 4, pp. 1489–1496, August 2005.

63. L. Masek and P. Kovesi, MATLAB Source Code for a Biometric Identification System Based on Iris Patterns, The School of Computer Science and Software Engineering, The University of Western Australia, 2003.

64. Center for Biometrics and Security Research, CASIA Iris Dataset, http://www.cbsr.ia. ac.cn.

65. A. J. Mansfield, J. L. Wayman, A. Dr, D. Rayner, and J. L. Wayman, Best practices in testing and reporting performance, Technical Report CMSC 14/02, Centre for Mathematics and Scientific Computing, National Physical Laboratory, Teddington, Middlesex, UK, 2002.

66. T. A. Louis, Confidence intervals for a binomial parameter after observing no successes, *The American Statistician*, vol. 35, no. 3, pp. 154–154, 1981.

67. B. D. Jovanovic and P. S. Levy, A look at the rule of three, *The American Statistician*, vol. 51, no. 2, pp. 137–139, 1997.

68. G. R. Doddington, M. A. Przybocki, A. F. Martin, and D. A. Reynolds, The NIST speaker recognition evaluation—overview, methodology, systems, results, perspective, *Speech Communication*, vol. 31, no. 2–3, pp. 225–254, 2000.

69. G. W. Snedecor and W. G. Cochran, *Statistical Methods,* 7th Edition. Iowa State University, Iowa City, 1980.

70. R. Bolle, N. Ratha, and S. Pankanti, Confidence interval measurement in performance analysis of biometrics systems using the bootstrap, in *Proceedings of the IEEE Workshop on Empirical Evaluation Methods in Computer Vision*, 2001.

71. N. Poh and S. Bengio, Estimating the confidence interval of expected performance curve in biometric authentication using joint bootstrap, in *Proceedings of the IEEE International Conference on Acoustics, Speech and Signal Processing*, vol. 2, April 2007, pp. 137–140.

72. R. M. Bolle, N. K. Ratha, and S. Pankanti, Error analysis of pattern recognition systems— The subsets bootstrap, *Computer Vision and Image Understanding*, vol. 93, no. 1, pp. 1–33, 2004.

73. R. Li, D. Tang, W. Li, and D. Zhang, Second-level partition for estimating FAR confidence intervals in biometric systems, in X. Jiang and N. Petkov, Eds., *Computer Analysis of Images and Patterns*, vol. 5702, Springer, Berlin, Heidelberg, 2009, pp. 58–65.

74. R. Li, B. Huang, R. Li, and W. Li, Test sample size determination for biometric systems based on confidence elasticity, in *Proceedings of the International Joint Conference on Neural Networks*, June 2012, pp. 1–7.

75. S. Dass, Y. Zhu, and A. Jain, Validating a biometric authentication system: Sample size requirements, *IEEE Transactions on Pattern Analysis and Machine Intelligence*, vol. 28, no. 12, pp. 1902–1319, 2006, pp. 1902–1913.

76. N. Poh, A. Martin, and S. Bengio, Performance generalization in biometric authentication using joint user-specific and sample bootstraps, *IEEE Transactions on Pattern Analysis and Machine Intelligence*, vol. 29, no. 3, pp. 492–498, 2007.

77. S. Cimato, M. Gamassi, V. Piuri, R. Sassi, and F. Scotti, Privacy-aware biometrics: Design and implementation of a multimodal verification system, in *Proceedings of the Annual Computer Security Applications Conference*, December 2008, pp. 130–139.

78. R. Donida Labati, V. Piuri, and F. Scotti, Agent-based image iris segmentation and multiple views boundary refining, in *Proceedings of the 3rd IEEE International Conference on Biometrics: Theory, Applications and Systems*, November 2009, pp. 1–7.

79. K. W. Bowyer, The results of the NICE II iris biometrics competition, *Pattern Recognition Letters*, vol. 33, no. 8, pp. 965–969, June 2012.

80. B. Kamgar-Parsi, W. Lawson, and B. Kamgar-Parsi, Toward development of a face recognition system for watchlist surveillance, *IEEE Transactions on Pattern Analysis and Machine Intelligence*, vol. 33, no. 10, pp. 1925–1937, October 2011.

81. J. Matey, O. Naroditsky, K. Hanna, R. Kolczynski, D. LoIacono, S. Mangru, M. Tinker, T. Zappia, and W. Zhao, Iris on the move: Acquisition of images for iris recognition in less constrained environments, *Proceedings of the IEEE*, vol. 94, no. 11, pp. 1936–1947, 2006.

82. K. Bowyer, K. Chang, and P. Flynn, A survey of approaches to three-dimensional face recognition, in *Proceedings of the 17th International Conference on Pattern Recognition*, vol. 1, August 2004, pp. 358–361.

83. M. Barni, T. Bianchi, D. Catalano, M. Di Raimondo, R. Donida Labati, P. Failla, D. Fiore, et al., A privacy-compliant fingerprint recognition system based on homomorphic encryption and fingercode templates, in *Proceedings of the 4th IEEE International Conference on Biometrics: Theory Applications and Systems*, September 2010, pp. 1–7.

84. M. Barni, T. Bianchi, D. Catalano, M. D. Raimondo, R. Donida Labati, P. Failla, D. Fiore, R. Lazzeretti, V. Piuri, F. Scotti, and A. Piva, Privacy-preserving fingercode authentication, in *Proceedings of the 12th ACM Workshop on Multimedia and Security*, ACM, New York, NY, September 2010, pp. 231–240.

85. T. Bianchi, R. Donida Labati, V. Piuri, A. Piva, F. Scotti, and S. Turchi, Implementing fingercode-based identity matching in the encrypted domain, in *Proceedings of the IEEE Workshop on Biometric Measurements and Systems for Security and Medical Applications*, September 2010, pp. 15–21.

86. S. Cimato, M. Gamassi, V. Piuri, R. Sassi, F. S. Cimato, and F. Scotti, A biometric verification system addressing privacy concerns, in *Proceedings of the International Conference on Computational Intelligence and Security*, December 2007, pp. 594–598.

87. K. Plataniotis, D. Hatzinakos, and J. Lee, ECG biometric recognition without fiducial detection, in *Proceedings of the Biometrics Symposium: Special Session on Research at the Biometric Consortium Conference*, August 2006, pp. 1–6.

88. R. Donida Labati, R. Sassi, and F. Scotti, ECG biometric recognition: Permanence analysis of qrs signals for 24 hours continuous authentication, in *Proceedings of the IEEE International Workshop on Information Forensics and Security*, November 2013.

89. A. Bonissi, R. Donida Labati, L. Perico, R. Sassi, F. Scotti, and L. Sparagino, A preliminary study on continuous authentication methods for photoplethysmographic biometrics, in *Proceedings of the 2013 IEEE Workshop on Biometric Measurements and Systems for Security and Medical Applications*, September 2013, pp. 28–33.

90. K. Ricanek, M. Savvides, D. Woodard, and G. Dozier, Unconstrained biometric identification: Emerging technologies, *Computer*, vol. 43, no. 2, pp. 56–62, 2010.

91. L. An, M. Kafai, and B. Bhanu, Dynamic Bayesian network for unconstrained face recognition in surveillance camera networks, *IEEE Journal on Emerging and Selected Topics in Circuits and Systems*, vol. 3, no. 2, pp. 155–164, 2013.

92. F. Wheeler, R. Weiss, and P. Tu, Face recognition at a distance system for surveillance applications, in *Proceedings of the 4th IEEE International Conference on Biometrics: Theory Applications and Systems*, 2010, pp. 1–8.

93. A. Amato, V. D. Lecce, and V. Piuri, *Semantic Analysis and Understanding of Human Behavior in Video Streaming*, Springer, New York, 2013.

94. B. Chen, J. Shen, and H. Sun, A fast face recognition system on mobile phone, in *Proceedings of the International Conference on Systems and Informatics*, May 2012, pp. 1783–1786.

95. P. Tresadern, T. Cootes, N. Poh, P. Matejka, A. Hadid, C. Levy, C. McCool, and S. Marcel, Mobile biometrics: Combined face and voice verification for a mobile platform, *IEEE Pervasive Computing*, vol. 12, no. 1, pp. 79–87, 2013.

96. A. Pentland and T. Choudhury, Face recognition for smart environments, *Computer*, vol. 33, no. 2, pp. 50–55, 2000.

97. T. Leyvand, C. Meekhof, Y.-C. Wei, J. Sun, and B. Guo, Kinect identity: Technology and experience, *Computer*, vol. 44, no. 4, pp. 94–96, 2011.

98. J. Y. Choi, W. De Neve, and Y. M. Ro, Towards an automatic face indexing system for actor-based video services in an IPTV environment, *IEEE Transactions on Consumer Electronics*, vol. 56, no. 1, pp. 147–155, 2010.

99. Y. Ishii, H. Hongo, K. Yamamoto, and Y. Niwa, Face and head detection for a real-time surveillance system, in *Proceedings of the 17th International Conference on Pattern Recognition*, vol. 3, August 2004, pp. 298–301.

100. S. Chin and K.-Y. Kim, Expressive 3D face for mobile devices, *IEEE Transactions on Consumer Electronics*, vol. 54, no. 3, pp. 1294–1302, 2008.

101. S. K. Zhou, R. Chellappa, and W. Zhao, *Unconstrained Face Recognition*. Springer, Boston, MA, 2006.
102. M. Turk and A. Pentland, Eigenfaces for recognition, *Journal of Cognitive Neuroscience*, vol. 3, no. 1, pp. 71–86, 1991.
103. R. Basri and D. Jacobs, Lambertian reflectance and linear subspaces, *IEEE Transactions on Pattern Analysis and Machine Intelligence*, vol. 25, no. 2, pp. 218–233, 2003.
104. J. J. Atick, P. A. Griffin, and A. N. Redlich, Statistical approach to shape from shading: Reconstruction of three-dimensional face surfaces from single two-dimensional images, *Neural Computation*, vol. 8, no. 6, pp. 1321–1340, 1996.
105. V. Blanz and T. Vetter, Face recognition based on fitting a 3D morphable model, *IEEE Transactions on Pattern Analysis and Machine Intelligence*, vol. 25, no. 9, pp. 1063–1074, 2003.
106. D. Yi, Z. Lei, and S. Z. Li, Towards pose robust face recognition, in *Proceedings of the IEEE Conference on Computer Vision and Pattern Recognition*, 2013, pp. 3539–3545.
107. X. Zhang and Y. Gao, Face recognition across pose: A review, *Pattern Recognition*, vol. 42, no. 11, pp. 2876–2896, 2009.
108. U. Park, Y. Tong, and A. Jain, Age-invariant face recognition, *IEEE Transactions on Pattern Analysis and Machine Intelligence*, vol. 32, no. 5, pp. 947–954, 2010.
109. H. Ling, S. Soatto, N. Ramanathan, and D. Jacobs, Face verification across age progression using discriminative methods, *IEEE Transactions on Information Forensics and Security*, vol. 5, no. 1, pp. 82–91, 2010.
110. S. Zhou, V. Krueger, and R. Chellappa, Probabilistic recognition of human faces from video, *Computer Vision and Image Understanding*, vol. 91, no. 1–2, pp. 214–245, 2003.
111. A. Scheenstra, A. Ruifrok, and R. C. Veltkamp, A survey of 3D face recognition methods, in *Lecture Notes in Computer Science*, Springer, New York, 2005, pp. 891–899.
112. Z. Sun, W. Dong, and T. Tan, Technology roadmap for smart iris recognition, in *Proceedings of the International Conference on Computer Graphics and Vision*, 2008, 12–19.
113. W. Dong, Z. Sun, T. Tan, and X. Qiu, Self-adaptive iris image acquisition system, in *Proceedings of SPIE*, 2008.
114. C. L. Fancourt, L. Bogoni, K. J. Hanna, Y. Guo, R. P. Wildes, N. Takahashi, and U. Jain, Iris recognition at a distance, in *Proceedings of Audio and Video Based Person Authentication*, 2005, pp. 1–13.
115. W. Dong, Z. Sun, and T. Tan, A design of iris recognition system at a distance, in *Proceedings of the Chinese Conference on Pattern Recognition*, November 2009, pp. 1–5.
116. R. Donida Labati, A. Genovese, V. Piuri, and F. Scotti, Iris segmentation: State of the art and innovative methods, in C. Liu and V. Mago, Eds., *Cross Disciplinary Biometric Systems*, vol. 37, Springer, Berlin, Germany, 2012, pp. 151–182.
117. H. Proença, Iris recognition: On the segmentation of degraded images acquired in the visible wavelength, *IEEE Transactions Pattern Analysis Machine Intelligence*, vol. 32, no. 8, pp. 1502–1516, 2010.
118. R. Donida Labati, V. Piuri, and F. Scotti, Neural-based iterative approach for iris detection in iris recognition systems, in *Proceedings of the IEEE Symposium on Computational Intelligence for Security and Defence Applications*, December 2009, pp. 1–6.
119. S. Shah and A. Ross, Iris segmentation using geodesic active contours, *IEEE Transactions on Information Forensics Security*, vol. 4, no. 4, pp. 824–836, 2009.
120. F. Scotti and V. Piuri, Adaptive reflection detection and location in iris biometric images by using computational intelligence techniques, *IEEE Transactions of Instrumentation and Measurement*, vol. 59, no. 7, pp. 1825–1833, 2010.

121. F. Scotti, Computational intelligence techniques for reflections identification in iris biometric images, in *Proceedings of the IEEE International Conference on Computational Intelligence for Measurement Systems and Applications*, June 2007, pp. 84–88.

122. Y.-H. Li and M. Savvides, An automatic iris occlusion estimation method based on high-dimensional density estimation, *IEEE Transactions Pattern Analysis and Machine Intelligence*, vol. 35, no. 4, pp. 784–796, 2013.

123. K. Nguyen, C. Fookes, S. Sridharan, and S. Denman, Quality-driven super-resolution for less constrained iris recognition at a distance and on the move, *IEEE Transactions on Information Forensics and Security*, vol. 6, no. 4, pp. 1248–1258, 2011.

124. M. D. Marsico, C. Galdi, M. Nappi, and D. Riccio, Firme: Face and iris recognition for mobile engagement, *Image and Vision Computing*, Vol. 32, pp. 1161–1172, 2014.

125. S. Schuckers, N. Schmid, A. Abhyankar, V. Dorairaj, C. Boyce, and L. Hornak, On techniques for angle compensation in nonideal iris recognition, *IEEE Transactions on Systems, Man, and Cybernetics, Part B: Cybernetics*, vol. 37, no. 5, pp. 1176–1190, 2007.

126. C.-T. Chou, S.-W. Shih, W.-S. Chen, V. Cheng, and D.-Y. Chen, Non-orthogonal view iris recognition system, *IEEE Transactions on Circuits and Systems for Video Technology*, vol. 20, no. 3, pp. 417–430, 2010.

127. D. Reid, M. Nixon, and S. Stevenage, Soft biometrics: Human identification using comparative descriptions, *IEEE Transactions on Pattern Analysis and Machine Intelligence*, vol. 36, no. 6, pp. 1216–1228, 2014.

128. K. Niinuma, U. Park, and A. Jain, Soft biometric traits for continuous user authentication, *IEEE Transactions on Information Forensics and Security*, vol. 5, no. 4, pp. 771–780, 2010.

129. Y. Ran, G. Rosenbush, and Q. Zheng, Computational approaches for real-time extraction of soft biometrics, in *Proceedings of the 19th International Conference on Pattern Recognition*, 2008, pp. 1–4.

130. J. Zhang, Y. Cheng, and C. Chen, Low resolution gait recognition with high frequency super resolution, in T.-B. Ho and Z.-H. Zhou, Eds., *PRICAI 2008: Trends in Artificial Intelligence*, vol. 5351, Springer, Berlin, Heidelberg, 2008, pp. 533–543.

131. Y. Guan, Y. Sun, C.-T. Li, and M. Tistarelli, Human gait identification from extremely low-quality videos: An enhanced classifier ensemble method, *IET Biometrics*, vol. 3, no. 2, pp. 84–93, 2014.

132. J. Bustard and M. Nixon, Toward unconstrained ear recognition from two-dimensional images, *IEEE Transactions on Systems, Man and Cybernetics, Part A: Systems and Humans*, vol. 40, no. 3, pp. 486–494, 2010.

133. R. Raposo, E. Hoyle, A. Peixinho, and H. Proenca, Ubear: A dataset of ear images captured on-the-move in uncontrolled conditions, in *Proceedings of the IEEE Workshop on Computational Intelligence in Biometrics and Identity Management*, April 2011, 84–90.

134. M. Ramalho, P. Correia, and L. Soares, Hand-based multimodal identification system with secure biometric template storage, *IET Computer Vision*, vol. 6, no. 3, pp. 165–173, 2012.

135. V. Kanhangad, A. Kumar, and D. Zhang, Contactless and pose invariant biometric identification using hand surface, *IEEE Transactions on Image Processing*, vol. 20, no. 5, pp. 1415–1424, 2011.

136. D. L. Woodard and P. J. Flynn, Finger surface as a biometric identifier, *Computer Vision and Image Understanding*, vol. 100, no. 3, pp. 357–384, 2005.

137. S. Ribaric and I. Fratric, A biometric identification system based on eigenpalm and eigenfinger features, *IEEE Transactions on Pattern Analysis and Machine Intelligence*, vol. 27, no. 11, pp. 1698–1709, 2005.

138. A. Genovese, V. Piuri, and F. Scotti, in S. Jajodia, Ed. *Touchless Palmprint Recognition Systems*. vol. 60, Springer, International Publishing, Cham, Swizerland, November 2014, p. 230.

139. A. Kumar, Incorporating cohort information for reliable palmprint authentication, in *Proceedings of the 6th Indian Conference on Computer Vision, Graphics Image Processing*, December 2008, pp. 583–590.

140. R.-X. Hu, W. Jia, D. Zhang, J. Gui, and L.-T. Song, Hand shape recognition based on coherent distance shape contexts, *Pattern Recognition*, vol. 45, no. 9, pp. 3348–3359, 2012.

141. Z. Zhang, S. Huang, Y. Xu, C. Chen, Y. Zhao, N. Gao, and Y. Xiao, 3D palmprint and hand imaging system based on full-field composite color sinusoidal fringe projection technique, *Applied Optics*, vol. 52, no. 25, pp. 6138–6145, 2013.

142. A. Kumar and D. Zhang, Personal recognition using hand shape and texture, *IEEE Transactions on Image Processing*, vol. 15, no. 8, pp. 2454–2461, 2006.

143. J. Doublet, O. Lepetit, and M. Revenu, Contactless hand recognition based on distribution estimation, in *Proceedings of the Biometrics Symposium*, September 2007, pp. 1–6.

144. A. de Santos Sierra, J. Casanova, C. Avila, and V. Vera, Silhouette-based hand recognition on mobile devices, in *Proceedings of the 43rd Annual International Carnahan Conference on Security Technology*, October 2009, pp. 160–166.

145. M. Wong, D. Zhang, W.-K. Kong, and G. Lu, Real-time palmprint acquisition system design, *IEEE Proceedings on Vision, Image and Signal Processing*, vol. 152, no. 5, pp. 527–534, October 2005.

146. V. Kanhangad, A Kumar, and D. Zhang, A unified framework for contactless hand verification, *IEEE Transactions on Information Forensics and Security*, vol. 6, no. 3, pp. 1014–1027, 2011.

147. J. D. Woodward Jr., N. M. Orlans, and P. T. Higgins, *Biometrics*. McGraw-Hill, Osborne, 2003.

148. A. Jain, A. Ross, and S. Pankanti, Biometrics: A tool for information security, *IEEE Transactions on Information Forensics and Security*, vol. 1, no. 2, pp. 125–143, 2006.

149. W. Babler, Embryologic development of epidermal ridges and their configurations, *Birth Defects Original Article Series*, vol. 27, no. 2, pp. 95–112, 1991.

150. A. K. Jain, S. Prabhakar, and S. Pankanti, On the similarity of identical twin fingerprints, *Pattern Recognition*, vol. 35, no. 11, pp. 2653–2663, 2002.

151. X. Tao, X. Chen, X. Yang, and J. Tian, Fingerprint recognition with identical twin fingerprints, *PLoS ONE*, vol. 7, no. 4, pp. e35704-1–e35704-7, 2012.

152. S. Pankanti, S. Prabhakar, and A. Jain, On the individuality of fingerprints, *IEEE Transactions on Pattern Analysis and Machine Intelligence*, vol. 24, no. 8, pp. 1010–1025, 2002.

153. K. Kryszczuk, P. Morier, and A. Drygajlo, Study of the distinctiveness of level 2 and level 3 features in fragmentary fingerprint comparison, in D. Maltoni and A. Jain, Eds., *Biometric Authentication*, vol. 3087, Springer, Berlin, Heidelberg, 2004, pp. 124–133.

154. M. Puertas, D. Ramos, J. Fierrez, J. Ortega-Garcia, and N. Exposito, Towards a better understanding of the performance of latent fingerprint recognition in realistic forensic conditions, in *Proceedings of the 20th International Conference on Pattern Recognition*, 2010, pp. 1638–1641.

155. N. K. Ratha and R. Bolle, *Automatic Fingerprint Recognition Systems,* Springer, New York, 2003.

156. T.-T. Truong, M.-T. Tran, and A.-D. Duong, Robust mobile device integration of a fingerprint biometric remote authentication scheme, in *Proceedings of the IEEE 26th International Conference on Advanced Information Networking and Applications*, March 2012, pp. 678–685.

157. S. B. Pan, D. Moon, Y. Gil, D. Ahn, and Y. Chung, An ultra-low memory fingerprint matching algorithm and its implementation on a 32-bit smart card, *IEEE Transactions on Consumer Electronics*, vol. 49, no. 2, pp. 453–459, 2003.

158. Z. Hou, W.-Y. Yau, and Y. Wang, A review on fingerprint orientation estimation, *Security and Communication Networks*, vol. 4, pp. 591–599, 2009.

159. A. Grasselli, On the automatic classification of fingerprints, in S. Watanabe, Ed., *Methodologies of Pattern Recognition*, Academic Press, New York, 1969, pp. 253–273.

160. A. Bazen and S. Gerez, Systematic methods for the computation of the directional fields and singular points of fingerprints, *IEEE Transactions on Pattern Analysis and Machine Intelligence*, vol. 24, no. 7, pp. 905–919, 2002.

161. Y. Mei, G. Cao, H. Sun, and R. Hou, A systematic gradient-based method for the computation of fingerprint's orientation field, *Computers and Electrical Engineering*, vol. 38, pp. 1035–1046, 2011.

162. L. Ji and Z. Yi, Fingerprint orientation field estimation using ridge projection, *Pattern Recognition*, vol. 41, no. 5, pp. 1491–1503, 2008.

163. T. Kamei, Image filter design for fingerprint enhancement, in N. Ratha and R. Bolle, Eds., *Automatic Fingerprint Recognition Systems*, Springer, New York, 2004, pp. 113–126.

164. S. Chikkerur, A. N. Cartwright, and V. Govindaraju, Fingerprint enhancement using STFT analysis, *Pattern Recognition*, vol. 40, no. 1, pp. 198–211, 2007.

165. M. Oliveira and N. Leite, A multiscale directional operator and morphological tools for reconnecting broken ridges in fingerprint images, *Pattern Recognition*, vol. 41, no. 1, pp. 367–377, 2008.

166. L. Min Liu and T.-S. Dai, A reliable fingerprint orientation estimation algorithm, *Journal of Information Science and Engineering*, vol. 27, no. 1, pp. 353–368, 2011.

167. B. Sherlock and D. Monro, A model for interpreting fingerprint topology, *Pattern Recognition*, vol. 26, no. 7, pp. 1047–1055, 1993.

168. L. Hong, Y. Wan, and A. Jain, Fingerprint image enhancement: Algorithm and performance evaluation, *IEEE Transactions on Pattern Analysis and Machine Intelligence*, vol. 20, no. 8, pp. 777–789, 1998.

169. X. Jiang, Fingerprint image ridge frequency estimation by higher order spectrum, in *Proceedings of the International Conference on Image Processing*, vol. 1, 2000, pp. 462–465.

170. A. Almansa and T. Lindeberg, Fingerprint enhancement by shape adaptation of scale-space operators with automatic scale selection, *IEEE Transactions on Image Processing*, vol. 9, no. 12, pp. 2027–2042, 2000.

171. C. Gottschlich, Curved-region-based ridge frequency estimation and curved gabor filters for fingerprint image enhancement, *IEEE Transactions on Image Processing*, vol. 21, no. 4, pp. 2220–2227, 2012.

172. M. Kawagoe and A. Tojo, Fingerprint pattern classification, *Pattern Recognition*, vol. 17, no. 3, pp. 295–303, 1984.

173. L. Fan, S. Wang, H. Wang, and T. Guo, Singular points detection based on zero-pole model in fingerprint images, *IEEE Transactions on Pattern Analysis and Machine Intelligence*, vol. 30, pp. 929–940, 2008.

174. J. Zhou, J. Gu, and D. Zhang, Singular points analysis in fingerprints based on topological structure and orientation field, in *Proceedings of the International Conference of Biometrics*, 2007, pp. 261–270.

175. H. Kekre and V. Bharadi, Fingerprint's core point detection using orientation field, in *Proceedings of the International Conference on Advances in Computing, Control, Telecommunication Technologies*, December 2009, pp. 150–152.

176. T. Liu, C. Zhang, and P. Hao, Fingerprint reference point detection based on local axial symmetry, in *Proceedings of the 18th International Conference on Pattern Recognition*, 2006, pp. 1050–1053.

177. J. L. A. Samatelo and E. O. T. Salles, Determination of the reference point of a fingerprint based on multiple levels of representation, in *Proceedings of the XXII Brazilian Symposium on Computer Graphics and Image Processing*, 2009, pp. 209–215.

178. H. Lam, Z. Hou, W. Yau, T. Chen, J. Li, and K. Sim, Reference point detection for arch type fingerprints, in *Proceedings of the 3rd International Conference on Advances in Biometrics*, 2009, pp. 666–674.

179. W.-C. Lin and R. C. Dubes, A review of ridge counting in dermatoglyphics, *Pattern Recognition*, vol. 16, no. 1, pp. 1–8, 1983.

180. A. Jain, S. Prabhakar, L. Hong, and S. Pankanti, Filterbank-based fingerprint matching, *IEEE Transactions on Image Processing*, vol. 9, no. 5, pp. 846–859, 2000.

181. R. Bansal, P. Sehgal, and P. Bedi, Minutiae extraction from fingerprint images—A review, *International Journal of Computer Science Issues*, vol. 8, no. 5, pp. 929–940, 2012.

182. R. C. Gonzalez and R. E. Woods, *Digital Image Processing*, 3rd Edition. Prentice-Hall, Upper Saddle River, NJ, 2006.

183. R. M. Stock, Automatic fingerprint reading, in *Proceedings of the Carnahan Conference on Electronic Crime Countermeasure*, 1972.

184. M. Verma, A. Majumdar, and B. Chatterjee, Edge detection in fingerprints, *Pattern Recognition*, vol. 20, no. 5, pp. 513–523, 1987.

185. J. Bartunek, M. Nilsson, J. Nordberg, and I. Claesson, Adaptive fingerprint binarization by frequency domain analysis, in *Proceedings of the 40th Asilomar Conference on Signals, Systems and Computers*, November 2006, pp. 598–602.

186. C. I. Watson, M. D. Garris, E. Tabassi, C. L. Wilson, R. M. Mccabe, S. Janet, and K. Ko, User's Guide to NIST Biometric Image Software (NBIS), online document published by the National Institute of Standards and Technology, https://www.nist.gov/customcf/get_pdf.cfm?pub_id=51097, 2007.

187. V. Espinosa-Duro, Fingerprints thinning algorithm, *IEEE Aerospace and Electronic Systems Magazine*, vol. 18, no. 9, pp. 28–30, 2003.

188. L. Ji, Z. Yi, L. Shang, and X. Pu, Binary fingerprint image thinning using template-based pcnns, *IEEE Transactions on Systems, Man, and Cybernetics, Part B: Cybernetics*, vol. 37, no. 5, pp. 1407–1413, 2007.

189. B. Fang, H. Wen, R.-Z. Liu, and Y.-Y. Tang, A new fingerprint thinning algorithm, in *Proceedings of the Chinese Conference on Pattern Recognition*, October 2010, pp. 1–4.

190. C. Arcelli and G. S. Di Baja, A width-independent fast thinning algorithm, *IEEE Transactions on Pattern Analysis and Machine Intelligence*, vol. 7, no. 4, pp. 463–474, 1985.

191. R. Bansal, P. Sehgal, and P. Bedi, Effective morphological extraction of true fingerprint minutiae based on the hit or miss transform, *International Journal of Biometrics and Bioinformatics*, vol. 4, pp. 471–485, 2010.

192. F. Zhao and X. Tang, Preprocessing and postprocessing for skeleton-based fingerprint minutiae extraction, *Pattern Recognition*, vol. 40, no. 4, pp. 1270–1281, 2007.

193. D. Maio and D. Maltoni, Neural network based minutiae filtering in fingerprints, in *Proceedings of the 14th International Conference on Pattern Recognition*, vol. 2, August 1998, pp. 1654–1658.

194. M. Gamassi, V. Piuri, and F. Scotti, Fingerprint local analysis for high-performance minutiae extraction, in *Proceedings of the IEEE International Conference on Image Processing*, vol. 3, September 2005, pp. 265–268.

195. S. Di Zenzo, L. Cinque, and S. Levialdi, Run-based algorithms for binary image analysis and processing, *IEEE Transactions on Pattern Analysis and Machine Intelligence*, vol. 18, no. 1, pp. 83–89, 1996.

196. Z. Shi and V. Govindaraju, A chaincode based scheme for fingerprint feature extraction, *Pattern Recognition Letters*, vol. 27, no. 5, pp. 462–468, 2006.

197. D. Maio and D. Maltoni, Direct gray-scale minutiae detection in fingerprints, *IEEE Transactions on Pattern Analysis and Machine Intelligence*, vol. 19, pp. 27–40, 1997.

198. J. Liu, Z. Huang, and K. L. Chan, Direct minutiae extraction from gray-level fingerprint image by relationship examination, in *Proceedings of the International Conference on Image Processing*, vol. 2, September 2000, pp. 427–430.

199. N. Canyellas, E. Cantó, G. Forte, and M. López, Hardware-software codesign of a fingerprint identification algorithm, in T. Kanade, A. Jain, and N. Ratha, Eds., *Audio- and Video-Based Biometric Person Authentication*, vol. 3546, Springer, Berlin, Heidelberg, 2005, pp. 165–200.

200. M.-T. Leung, W. Engeler, and P. Frank, Fingerprint image processing using neural networks, in *Proceedings of the IEEE Region 10 Conference on Computer and Communication Systems*, vol. 2, September 1990, pp. 582–586.

201. V. Sagai and A. Koh Jit Beng, Fingerprint feature extraction by fuzzy logic and neural networks, in *Proceedings of the International Conference on Neural Information Processing*, vol. 3, 1999, pp. 1138–1142.

202. K. Nilsson and J. Bigun, Using linear symmetry features as a pre-processing step for fingerprint images, in *Proceedings of the 3rd International Conference on Audio and Video Based Person Authentication*, 2001, pp. 247–252.

203. J. D. Stoszlisa and A. Alyea, Automated system for fingerprint authentication using pores and ridge structure, in *Proceedings of SPIE (Automatic Systems for the Identification and Inspection of Humans)*, vol. 2277, 1994, pp. 210–223.

204. A. K. Jain, Y. Chen, and M. Demirkus, Pores and ridges: High-resolution fingerprint matching using level 3 features, *IEEE Transactions on Pattern Analysis and Machine Intelligence*, vol. 29, pp. 15–27, 2007.

205. G. Marcialis, F. Roli, and A. Tidu, Analysis of fingerprint pores for vitality detection, in *Proceedings of the 20th International Conference on Pattern Recognition*, August 2010, pp. 1289–1292.

206. Federal Bureau of Investigation, The Science of Fingerprints. Classification and Uses, 2006.

207. J. Berry and D. Stoney, The history and development of fingerprinting, in H. Lee and R. Gaensslen, Eds., *Advances in Fingerprint Technology*, CRC Press, Boca Raton, FL, 2001, pp. 1–40.

208. S. Memon, M. Sepasian, and W. Balachandran, Review of finger print sensing technologies, in *Proceedings of the IEEE International Multitopic Conference*, December 2008, pp. 226–231.

209. Federal Bureau of Investigation, Electronic fingerprint transmission specification. http://nla.gov.au/nla.cat-vn4185009.

210. Federal Bureau of Investigation, http://www.fbi.gov.

211. J. N. Bradley, C. M. Brislawn, and T. Hopper, FBI wavelet/scalar quantization standard for gray-scale fingerprint image compression, in *Proceedings of SPIE*, vol. 1961, 1993, pp. 293–304.

212. A. Skodras, C. Christopoulos, and T. Ebrahimi, The JPEG 2000 still image compression standard, *IEEE Signal Processing Magazine*, vol. 18, no. 5, pp. 36–58, 2001.

213. N. M. Allinson, Fingerprint compression, in S. Z. Li and A. K. Jain, Eds., *Encyclopedia of Biometrics*, 2009, pp. 447–452.

214. F. Alonso-Fernandez, J. Fierrez, J. Ortega-Garcia, J. Gonzalez-Rodriguez, H. Fronthaler, K. Kollreider, and J. Bigun, A comparative study of fingerprint image-quality estimation methods, *IEEE Transactions on Information Forensics and Security*, vol. 2, no. 4, pp. 734–743, 2007.

215. R. Stewart, M. Estevao, and A Adler, Fingerprint recognition performance in rugged outdoors and cold weather conditions, in *Proceedings of the 3rd IEEE International Conference on Biometrics: Theory, Applications, and Systems*, September 2009, pp. 1–6.

216. S. Lee, H. seung Choi, K. Choi, and J. Kim, Fingerprint-quality index using gradient components, *IEEE Transactions on Information Forensics and Security*, vol. 3, no. 4, pp. 792–800, 2008.

217. L. Shen, A. C. Kot, and W. M. Koo, Quality measures of fingerprint images, in *Proceedings of the 3rd International Conference on Audio and Video Based Biometric Person Authentication*, 2001, pp. 266–271.

218. N. Ratha and R. Bolle, Fingerprint image quality estimation, IBM Computer Science Research, Report RC21622, 1999.

219. D. Yu, L. Ma, H. Lu, and Z. Chen, Fusion method of fingerprint quality evaluation: From the local gabor feature to the global spatial-frequency structures, in *Proceedings of the 8th International Conference on Advanced Concepts For Intelligent Vision Systems*, 2006, pp. 776–785.

220. X. Yang and Y. Luo, A classification method of fingerprint quality based on neural network, in *Proceedings of the International Conference on Multimedia Technology*, July 2011, pp. 20–23.

221. E. Tabassi, C. Wilson, and C. Watson, Fingerprint image quality, National Institute of Standards and Technology, Technical Report NISTIR 7151, August 2004.

222. S. Greenberg, M. Aladjem, D. Kogan, and I. Dimitrov, Fingerprint image enhancement using filtering techniques, in *Proceedings of the International Conference on Pattern Recognition*, vol. 3, 2000, pp. 322–325.

223. L. O'Gorman and J. V. Nickerson, An approach to fingerprint filter design, *Pattern Recognition*, vol. 22, no. 1, pp. 29–38, 1989.

224. W. Wang, J. Li, F. Huang, and H. Feng, Design and implementation of Log-Gabor filter in fingerprint image enhancement, *Pattern Recognition Letters*, vol. 29, no. 3, pp. 301–308, 2008.

225. M. Ghafoor, I. Taj, W. Ahmad, and N. Jafri, Efficient 2-fold contextual filtering approach for fingerprint enhancement, *IET Image Processing*, vol. 8, no. 7, pp. 417–425, 2014.

226. H. Fronthaler, K. Kollreider, and J. Bigun, Local features for enhancement and minutiae extraction in fingerprints, *IEEE Transactions on Image Processing*, vol. 17, no. 3, pp. 354–363, 2008.

227. J. Zhou, F. Chen, N. Wu, and C. Wu, Crease detection from fingerprint images and its applications in elderly people, *Pattern Recognition*, vol. 42, no. 5, pp. 896–906, 2009.

228. T. Hatano, T. Adachi, S. Shigematsu, H. Morimura, S. Onishi, Y. Okazaki, and H. Kyuragi, A fingerprint verification algorithm using the differential matching rate, in *Proceedings of the International Conference on Pattern Recognition*, vol. 3, 2002, pp. 799–802.

229. K. Nandakumar and A. K. Jain, Local correlation-based fingerprint matching, in *Proceedings of the Indian Conference on Computer Vision, Graphics and Image Processing*, 2004, pp. 503–508.

230. N. Yager and A. Amin, Fingerprint verification based on minutiae features: A review, *Pattern Analysis and Applications*, vol. 7, pp. 94–113, 2004.

231. A. Bishnu, S. Das, S. C. Nandy, and B. B. Bhattacharya, Simple algorithms for partial point set pattern matching under rigid motion, *Pattern Recognition*, vol. 39, no. 9, pp. 1662–1671, 2006.

232. N. Ratha, K. Karu, S. Chen, and A. Jain, A real-time matching system for large fingerprint databases, *IEEE Transactions on Pattern Analysis and Machine Intelligence*, vol. 18, no. 8, pp. 799–813, 1996.

233. S. Ranade and A. Rosenfeld, Point pattern matching by relaxation, *Pattern Recognition*, vol. 12, no. 4, pp. 269–275, 1980.

234. W. Sheng, G. Howells, M. Fairhurst, and F. Deravi, A memetic fingerprint matching algorithm, *IEEE Transactions on Information Forensics and Security*, vol. 2, no. 3, pp. 402–412, 2007.

235. H. Lam, W. Yau, T. Chen, Z. Hou, and H. Wang, Fingerprint pre-alignment for hybrid match-on-card system, in *Proceedings of the 6th International Conference on Information, Communications Signal Processing*, December 2007, pp. 1–4.

236. X. Jiang and W.-Y. Yau, Fingerprint minutiae matching based on the local and global structures, in *Proceedings of the 5th International Conference on Pattern Recognition*, vol. 2, 2000, pp. 1038–1041.

237. N. Ratha, R. Bolle, V. Pandit, and V. Vaish, Robust fingerprint authentication using local structural similarity, in *Proceedings of the Workshop on Applications of Computer Vision*, January 2000, pp. 29–34.

238. G. Bebis, T. Deaconu, and M. Georgiopoulos, Fingerprint identification using delaunay triangulation, in *Proceedings of the IEEE International Conference on Intelligence, Information, and Systems*, 1999, pp. 452–459.

239. G. Parziale and A. Niel, Fingerprint matching using minutiae triangulation, in *Proceedings of the International Conference on Biometric Authentication*, vol. 3072, July 2004, pp. 241–248.

240. X. Liang, A. Bishnu, and T. Asano, A robust fingerprint indexing scheme using minutia neighborhood structure and low-order delaunay triangles, *IEEE Transactions on Information Forensics and Security*, vol. 2, no. 4, pp. 721–733, 2007.

241. X. Tong, S. Liu, J. Huang, and X. Tang, Local relative location error descriptor-based fingerprint minutiae matching, *Pattern Recognition Letters*, vol. 29, no. 3, pp. 286–294, 2008.

242. L. Sha, F. Zhao, and X. Tang, Minutiae-based fingerprint matching using subset combination, in *Proceedings of the International Conference on Pattern Recognition*, vol. 4, 2006, pp. 566–569.

243. A. Ross and R. Nadgir, A thin-plate spline calibration model for fingerprint sensor interoperability, *IEEE Transactions on Knowledge and Data Engineering*, vol. 20, no. 8, pp. 1097–1110, 2008.

244. R. Cappelli, D. Maio, and D. Maltoni, Modelling plastic distortion in fingerprint images, in *Advances in Pattern Recognition*, vol. 2013, 2001, pp. 371–378.

245. L. Coetzee and E. C. Botha, Fingerprint recognition in low quality images, *Pattern Recognition*, vol. 26, no. 10, pp. 1441–1460, 1993.

246. R. Zhou, S. Sin, D. Li, T. Isshiki, and H. Kunieda, Adaptive sift-based algorithm for specific fingerprint verification, in *Proceedings of the International Conference on Hand-Based Biometrics*, November 2011, pp. 1–6.

247. J. Feng and A. Cai, Fingerprint representation and matching in ridge coordinate system, in *Proceedings of the International Conference on Pattern Recognition*, vol. 4, 2006, 485–488.

248. Q. Zhao, L. Zhang, D. Zhang, and N. Luo, Direct pore matching for fingerprint recognition, in *Proceedings of the 3rd International Conference on Advances in Biometrics*, 2009, pp. 597–606.

249. A. Jain, Y. Chen, and M. Demirkus, Pores and ridges: Fingerprint matching using level 3 features, in *Proceedings of the International Conference on Pattern Recognition*, vol. 4, 2006, pp. 477–480.

250. R. Cappelli and D. Maio, The state of the art in fingerprint classification, in N. Ratha and R. Bolle, Eds., *Automatic Fingerprint Recognition Systems*, Springer, New York, 2004, pp. 183–205.

251. G. A. Drets and H. G. Liljenström, Fingerprint sub-classification: A neural network approach, in *Intelligent Biometric Techniques in Fingerprint and Face Recognition*, CRC Press, Boca Raton, FL, 1999, pp. 107–134.

252. J. Araque, M. Baena, B. Chalela, D. Navarro, and P. Vizcaya, Synthesis of fingerprint images, in *Proceedings of the 16th International Conference on Pattern Recognition*, vol. 2, 2002, pp. 422–425.

253. M. Kücken and A. C. Newell, A model for fingerprint formation, *Europhysics Letters*, vol. 68, no. 1, p. 141, 2004.

254. M. Kücken, Models for fingerprint pattern formation, *Forensic Science International*, vol. 171, no. 2–3, pp. 85–96, 2007.

255. U.-K. Cho, J.-H. Hong, and S.-B. Cho, Automatic fingerprints image generation using evolutionary algorithm, in *Proceedings of the 20th International Conference on Industrial, Engineering, and Other Applications of Applied Intelligent Systems*, 2007, pp. 444–453.

256. R. Cappelli, Synthetic fingerprint generation, in A. J. D. Maltoni, D. Maio, and S. Prabhakar, Eds., *Handbook of Fingerprint Recognition,* 2nd Edition. Springer, London, 2009, pp. 271–301.

257. J. Xue, S. Xing, Y. Guo, and Z. Liu, Fingerprint generation method based on gabor filter, in *Proceedings of the International Conference on Computer Application and System Modeling*, vol. 8, October 2010, pp. 115–119.

258. P. Johnson, F. Hua, and S. Schuckers, Texture modeling for synthetic fingerprint generation, in *Proceedings of the IEEE Conference on Computer Vision and Pattern Recognition Workshops*, June 2013, pp. 154–159.

259. G. Parziale, Touchless fingerprinting technology, in N. K. Ratha and V. Govindaraju, Eds., *Advances in Biometrics*, Springer, London, 2008, pp. 25–48.

260. R. Donida Labati, A. Genovese, V. Piuri, and F. Scotti, Touchless fingerprint biometrics: A survey on 2D and 3D technologies, *Journal of Internet Technology*, vol. 15, no. 3, pp. 325–332, 2014, 1607-9264.

261. R. Donida Labati, A. Genovese, V. Piuri, and F. Scotti, Virtual environment for 3-D synthetic fingerprints, in *Proceedings of the IEEE International Conference on Virtual Environments, Human-Computer Interfaces and Measurement Systems*, July 2012, 48–53.

262. V. Piuri and F. Scotti, Fingerprint biometrics via low-cost sensors and webcams, in *Proceedings of the 2nd IEEE International Conference on Biometrics: Theory, Applications and Systems*, October 2008, pp. 1–6.

263. AFIS and Biometrics Consulting, http://www.afisandbiometrics.com.

264. M. O. Derawi, B. Yang, and C. Busch, Fingerprint recognition with embedded cameras on mobile phone, in *Proceedings of the 3rd International ICST Conference MobiSec*, vol. 94, 2011, pp. 136–147.

265. F. Han, J. Hu, M. Alkhathami, and K. Xi, Compatibility of photographed images with touch-based fingerprint verification software, in *Proceedings of the IEEE Conference on Industrial Electronics and Applications*, June 2011, pp. 1034–1039.

266. B. Hiew, A. Teoh, and D. Ngo, Automatic digital camera based fingerprint image preprocessing, in *Proceedings of the International Conference on Computer Graphics, Imaging and Visualisation*, July 2006, pp. 182–189.

267. B. Hiew, B. Andrew, and Y. Pang, Digital camera based fingerprint recognition, in *Proceedings of the IEEE International Conference on Telecommunications and Malaysia International Conference on Communications*, May 2007, pp. 676–681.

268. B. Hiew, A. Teoh, and D. Ngo, Preprocessing of fingerprint images captured with a digital camera, in *Proceedings of the International Conference on Control, Automation, Robotics and Vision*, December 2006, pp. 1–6.

269. Y. Song, C. Lee, and J. Kim, A new scheme for touchless fingerprint recognition system, in *Proceedings of the International Symposium on Intelligent Signal Processing and Communication Systems*, November 2004, pp. 524–527.

270. L. Wang, R. H. A. El-Maksoud, J. M. Sasian, and V. S. Valencia, in R. J. Koshel and G. G. Gregory, Eds., *Illumination Scheme for High-Contrast, Contactless Fingerprint Images*, in Proceedings of SPIE, pp. 742911-1–742911-7, vol. 7429, 2009.

271. C. Lee, S. Lee, and J. Kim, A study of touchless fingerprint recognition system, in D.-Y. Yeung, J. Kwok, A. Fred, F. Roli, and D. de Ridder, Eds., *Structural, Syntactic, and Statistical Pattern Recognition*, vol. 4109, Springer, Berlin, Heidelberg, 2006, pp. 358–365.

272. L. Wang, R. H. A. El-Maksoud, J. M. Sasian, W. P. Kuhn, K. Gee, and V. S. Valencia, in R. J. Koshel and G. G. Gregory, Eds., *A Novel Contactless Aliveness-Testing (CAT) Fingerprint*, in Proceedings of SPIE, pp. 742915-1–742915-11, vol. 7429, 2009.

273. E. Sano, T. Maeda, T. Nakamura, M. Shikai, K. Sakata, M. Matsushita, and K. Sasakawa, Fingerprint authentication device based on optical characteristics inside a finger, in *Proceedings of the Computer Vision and Pattern Recognition Workshop*, June 2006, pp. 27–32.

274. H. Choi, K. Choi, and J. Kim, Mosaicing touchless and mirror-reflected fingerprint images, *IEEE Transactions on Information Forensics and Security*, vol. 5, no. 1, pp. 52–61, March 2010.

275. F. Liu, D. Zhang, C. Song, and G. Lu, Touchless multiview fingerprint acquisition and mosaicking, *IEEE Transactions on Instrumentation and Measurement*, vol. 62, no. 9, pp. 2492–2502, 2013.

276. D. Noh, H. Choi, and J. Kim, Touchless sensor capturing five fingerprint images by one rotating camera, *Optical Engineering*, vol. 50, no. 11, pp. 113202–113212, 2011.

277. N. Francisco, A. Zaghetto, B. Macchiavello, E. da Silva, M. Lima-Marques, N. Rodrigues, and S. de Faria, Compression of touchless multiview fingerprints, in *Proceedings of the IEEE Workshop on Biometric Measurements and Systems for Security and Medical Applications*, September 2011, pp. 1–5.

278. C. Lee, S. Lee, J. Kim, and S.-J. Kim, Preprocessing of a fingerprint image captured with a mobile camera, in D. Zhang and A. Jain, Eds., *Advances in Biometrics*, vol. 3832, Springer, Berlin, Heidelberg, 2005, pp. 348–355.

279. M. Khalil and F. Wan, A review of fingerprint pre-processing using a mobile phone, in *Proceedings of the Conference on Wavelet Analysis and Pattern Recognition*, 2012, 152–157.

280. G. Parziale and Y. Chen, Advanced technologies for touchless fingerprint recognition, in M. Tistarelli, S. Z. Li, and R. Chellappa, Eds., *Handbook of Remote Biometrics*, Springer, London, 2009, pp. 83–109.

281. B. Y. Hiew, A. B. J. Teoh, and O. S. Yin, A secure digital camera based fingerprint verification system, *Journal of Visual Communication and Image Representation*, vol. 21, no. 3, pp. 219–231, 2010.

282. A. Kumar and Y. Zhou, Contactless fingerprint identification using level zero features, in *Proceedings of the IEEE Conference on Computer Vision and Pattern Recognition Workshops*, June 2011, pp. 114–119.

283. A Kumar and Y. Zhou, Human identification using finger images, *IEEE Transactions on Image Processing*, vol. 21, no. 4, pp. 2228–2244, April 2012.

284. S. Mil'shtein, M. Baier, C. Granz, and P. Bustos, Mobile system for fingerprinting and mapping of blood—vessels across a finger, in *Proceedings of the IEEE Conference on Technologies for Homeland Security*, May 2009, pp. 30–34.

285. G. Parziale, E. Diaz-Santana, and R. Hauke, The Surround Imager: A multi-camera touchless device to acquire 3D rolled-equivalent fingerprints, in *Proceedings of the International Conference on Biometrics*, 2006, pp. 244–250.

286. F. Liu and D. Zhang, 3D fingerprint reconstruction system using feature correspondences and prior estimated finger model, *Pattern Recognition*, vol. 47, no. 1, pp. 178–193, 2014.

287. G. Abramovich, K. Harding, Q. Hu, S. Manickam, M. Ganesh, and C. Nafis, Method and system for contactless fingerprint detection and verification, US Patent App. 12/694,840, March 2011.

288. Y. Wang, L. Hassebrook, and D. Lau, Data acquisition and processing of 3-D fingerprints, *IEEE Transactions on Information Forensics and Security*, vol. 5, no. 4, pp. 750–760, 2010.

289. Y. Wang, L. G. Hassebrook, and D. L. Lau, Noncontact, depth-detailed 3D fingerprinting, *SPIE Newsroom*, November 2009.

290. S. Huang, Z. Zhang, Y. Zhao, J. Dai, C. Chen, Y. Xu, E. Zhang, and L. Xie, 3D fingerprint imaging system based on full-field fringe projection profilometry, *Optics and Lasers in Engineering*, vol. 52, pp. 123–130, 2014.

291. W. Xie, Z. Song, and X. Zhang, A novel photometric method for real-time 3D reconstruction of fingerprint, in G. Bebis, R. Boyle, B. Parvin, D. Koracin, R. Chung, R. Hammound, M. Hussain et al., *Advances in Visual Computing*, vol. 6454, Springer, Berlin, Germany, 2010, pp. 31–40.

292. W. Xie, Z. Song, and R. Chung, Real-time three-dimensional fingerprint acquisition via a new photometric stereo means, *Optical Engineering*, vol. 52, no. 10, pp. 103103-1–103103-10, 2013.

293. A Baradarani, R. Maev, and F. Severin, Resonance based analysis of acoustic waves for 3D deep-layer fingerprint reconstruction, in *Proceedings of the IEEE International Ultrasonics Symposium*, July 2013, pp. 713–716.

294. A. Kumar and C. Kwong, Towards contactless, low-cost and accurate 3D fingerprint identification, in *Proceedings of the IEEE Conference on Computer Vision and Pattern Recognition*, June 2013, pp. 3438–3443.

295. Y. Wang, D. L. Lau, and L. G. Hassebrook, Fit-sphere unwrapping and performance analysis of 3D fingerprints, *Applied Optics*, vol. 49, no. 4, pp. 592–600, 2010.

296. S. Shafaei, T. Inanc, and L. Hassebrook, A new approach to unwrap a 3-D fingerprint to a 2-D rolled equivalent fingerprint, in *Proceedings of the 3rd IEEE International Conference on Biometrics: Theory, Applications, and Systems*, September 2009, pp. 1–5.

297. Q. Zhao, A. Jain, and G. Abramovich, 3D to 2D fingerprints: Unrolling and distortion correction, in *Proceedings of the International Joint Conference on Biometrics*, October 2011, pp. 1–8.

298. S. S. Arora, K. Cao, A. K. Jain, and N. G. Paulter Jr., 3D fingerprint phantoms, in *Proceedings of the 22nd International Conference on Pattern Recognition*, August 2014.

299. F. Bellocchio, N. Borghese, S. Ferrari, and V. Piuri, *3D Surface Reconstruction: Multi-Scale Hierarchical Approaches*, Springer, New York, 2013.

300. G. Paar, M. d. Perucha, A. Bauer, and B. Nauschnegg, Photogrammetric fingerprint unwrapping, *Journal of Applied Geodesy*, vol. 2, pp. 13–20, 2008.
301. R. I. Hartley and A. Zisserman, *Multiple View Geometry in Computer Vision,* 2nd Edition. Cambridge University Press, Cambridge, UK, 2004.
302. Y. Chen, G. Parziale, E. Diaz-Santana, and A. Jain, 3D touchless fingerprints: Compatibility with legacy rolled images, in *Proceedings of the Biometrics Symposium: Special Session on Research at the Biometric Consortium Conference*, September 2006, pp. 1–6.
303. R. Donida Labati, A. Genovese, V. Piuri, and F. Scotti, Two-view contactless fingerprint acquisition systems: A case study for clay artworks, in *Proceedings of the IEEE Workshop on Biometric Measurements and Systems for Security and Medical Applications*, September 2012.
304. M. J. Brooks, in B. K. P. Horn, Ed. *Shape from Shading*, MIT Press, Cambridge, USA, 1989.
305. R. J. Woodham, Photometric method for determining surface orientation from multiple images, *Optical Engineering*, vol. 19, no. 1, pp. 191139-1–191139-6, 1980.
306. D. Koller, L. Walchshäusl, G. Eggers, F. Neudel, U. Kursawe, P. Kühmstedt, M. Heinze, et al., 3D capturing of fingerprints—On the way to a contactless certified sensor, in *Proceedings of the Special Interest Group on Biometrics and Electronic Signatures*, 2011, pp. 33–44.
307. X. Pang, Z. Song, and W. Xie, Extracting valley-ridge lines from point-cloud-based 3D fingerprint models, *IEEE Computer Graphics and Applications*, vol. 33, no. 4, pp. 73–81, 2013.
308. C. Dorai and A. K. Jain, COSMOS-A representation scheme for 3D free-form objects, *IEEE Transactions on Pattern Analysis and Machine Intelligence*, vol. 19, no. 10, pp. 1115–1130, 1997.
309. Touchless Biometric Systems, TBS, http://www.tbs-biometrics.com.
310. Touchless Sensor Technology, TST, http://www.tst-biometrics.com.
311. Safran Morpho, http://www.morpho.com.
312. Mitsubishi Electric, http://www.mitsubishielectric.com.
313. S. Milshtein, A. Pillai, V. O. Kunnil, M. Baier, and P. Bustos, Applications of contactless fingerprinting, in J. Yang and N. Poh, Eds., *Recent Application in Biometrics*, InTech, 2011, pp. 107–134.
314. R. Donida Labati, V. Piuri, and F. Scotti, Neural-based quality measurement of fingerprint images in contactless biometric systems, in *Proceedings of the International Joint Conference on Neural Networks*, July 2010, pp. 1–8.
315. R. Kohavi and G. H. John, Wrappers for feature subset selection, *Artificial Intelligence*, vol. 97, no. 1–2, pp. 273–324, 1997.
316. I. Guyon and A. Elisseeff, An introduction to variable and feature selection, *Journal of Machine Learning Research*, vol. 3, pp. 1157–1182, 2003.
317. C. Alippi, P. Braione, V. Piuri, and F. Scotti, A methodological approach to multi-sensor classification for innovative laser material processing units, vol. 3, 2001, 1762–1767.
318. R. Donida Labati, A. Genovese, V. Piuri, and F. Scotti, Measurement of the principal singular point in contact and contactless fingerprint images by using computational intelligence techniques, in *Proceedings of the IEEE International Conference on Computational Intelligence for Measurement Systems and Applications*, September 2010, 18–23.
319. R. Donida Labati, A. Genovese, V. Piuri, and F. Scotti, Contactless fingerprint recognition: A neural approach for perspective and rotation effects reduction, in *Proceedings of the IEEE*

Workshop on Computational Intelligence in Biometrics and Identity Management, April 2013, pp. 22–30.

320. R. Donida Labati, V. Piuri, and F. Scotti, A neural-based minutiae pair identification method for touchless fingerprint images, in *Proceedings of the IEEE Workshop on Computational Intelligence in Biometrics and Identity Management*, April 2011.

321. D. G. Lowe, Distinctive image features from scale-invariant keypoints, in *Proceedings of the International Joint Conference on Computer Vision*, vol. 60, November 2004, 91–110.

322. Z. Zhang, A flexible new technique for camera calibration, *IEEE Transactions on Pattern Analysis and Machine Intelligence*, vol. 22, no. 11, pp. 1330–1334, 2000.

323. J. Heikkila and O. Silvén, A four-step camera calibration procedure with implicit image correction, *Proceedings of the IEEE Conference on Computer Vision and Pattern Recognition*, pp. 1106–1112, 1997.

324. R. Donida Labati, A. Genovese, V. Piuri, and F. Scotti, Fast 3-D fingertip reconstruction using a single two-view structured light acquisition, in *Proceedings of the IEEE Workshop on Biometric Measurements and Systems for Security and Medical Applications*, September 2011, pp. 1–8.

325. M. Brown, D. Burschka, and G. Hager, Advances in computational stereo, *IEEE Transactions on Pattern Analysis and Machine Intelligence*, vol. 25, no. 8, pp. 993–1008, August 2003.

326. P. D. Kovesi, MATLAB and Octave functions for computer vision and image processing, Centre for Exploration Targeting, School of Earth and Environment, The University of Western Australia. http://www.csse.uwa.edu.au/~pk/research/matlabfns.

327. R. Donida Labati, A. Genovese, V. Piuri, and F. Scotti, Quality measurement of unwrapped three-dimensional fingerprints: A neural networks approach, in *Proceedings of the International Joint Conference on Neural Networks*, June 2012, 1123–1130.

328. R. Donida Labati, A. Genovese, V. Piuri, and F. Scotti, Accurate 3D fingerprint virtual environment for biometric technology evaluations and experiment design, in *Proceedings of the IEEE International Conference on Computational Intelligence and Virtual Environments for Measurement Systems and Applications*, July 2013, pp. 43–48.

329. Blender, http://www.blender.org.

330. A. K. Jain, R. P. Duin, and J. Mao, Statistical pattern recognition: A review, *IEEE Transactions on Pattern Analysis and Machine Intelligence*, vol. 22, pp. 4–37, 2000.

331. R. O. Duda, P. E. Hart, and D. G. Stork, *Pattern Classification,* 2nd Edition. Wiley-Interscience, Hoboken, New Jersey, 2000.

332. Neurotechnology, http://www.neurotechnology.com.

333. CROSSMATCH Technologies, http://www.neurotechnology.com.

334. R. Donida Labati, V. Piuri, and F. Scotti, Measurement of the principal singular point in fingerprint images: A neural approach, in *Proceedings of the IEEE International Conference on Computational Intelligence for Measurement Systems and Applications*, September 2010, 18–23.

335. R. Guerchouche and F. Coldefy, Camera calibration methods evaluation procedure for images rectification and 3D reconstruction, Orange Labs, France Telecom R & D, 2008.

336. A. Ross and A. K. Jain, Multimodal Biometrics: An overview, in *Proceedings of 12th Conference European on Signal Processing*, 2004, pp. 1221–1224.

337. NVIDIA CUDA Compute Unified Device Architecture—Programming Guide, https://developer.nvidia.com, 2007.

338. International Organization for Standards, ISO 9241-11 Ergonomic Requirements for Office Work with Visual Display Terminals (VDTs)—Part 11: Guidance on Usability, 1998.

339. B. R. Hunt, R. L. Lipsman, and J. M. Rosenberg, *A Guide to MATLAB: For Beginners and Experienced Users*. Cambridge University Press, New York, NY, 2001.

340. M. Theofanos, B. Stanton, C. Sheppard, R. Micheals, and N. Zhang, Usability testing of ten-print fingerprint capture, NISTIR, Technical Report, 2007.

Index

Printed and bound by CPI Group (UK) Ltd, Croydon, CR0 4YY

23/10/2024

01777674-0008